Al  S

JODY CALL
HANDBOOK
VOLUME I

Compiled by
Sandee Shaffer Johnson

Edited by
Robert W. Morgan

TALISMAN BOOKS

Copyright © 1994 by The Talisman Media Group, Inc.
All Rights Reserved

Published by Talisman Books
A division of The Talisman Media Group, Inc.
336 Bon Air Center, Suite 341
Greenbrae, California 94904
(415) 461-6409

For further information please write
or call/fax (415) 461-6409

DISCLAIMER

Every diligent and reasonable effort has been made by the compiler to locate and secure permission for the inclusion of all copyrighted material appearing in this book. Due to the very transient nature of military cadences, most are passed along verbally; hence their true authorship may never be noted. In this and its companion works, many of the contributors are precisely that: as contributors they did not author the chants themselves but are generously "passing them along" to be used and remembered by succeeding generations of military personnel. However, if any such acknowledgements have been inadvertently omitted, the compiler, editor, and the publisher would deeply appreciate receiving full and corroborative information so that proper credit can be given in future editions.

Library of Congress Cataloging-in-Publication Data
The New American Cadences Jody Call Handbook /
compiled by Sandee Shaffer Johnson; edited by Robert W. Morgan.
Without the music; in part with indications of tunes.

ISBN 1-885969-00-7 (v.1: paper): $6.95
ISBN 1-885969-01-5 (v.2: paper): $6.95

1. Jody Calls - texts. 2. War songs - United States - Texts. 3. Folk Songs, English - United States - Texts.
I. Johnson, Sandee Shaffer, 1951 - . II. Morgan, Robert W. 1935 - .
III. Title: Jody Call Handbook.

Printed in the United States of America
10 9 8 7 6 5 4 3

To the American Soldier, Sailor, Airman, and Marine —

Past, Present, and Future

Contents

Acknowledgements 7

Introduction 11

CHAPTER ONE
- *A Brief History of Military Cadences* 17

CHAPTER TWO
- *Assorted Chants for Mixing and Creating* 31

CHAPTER THREE
- *General Cadences* 53

CHAPTER FOUR
- *Competition and Rivalry* 95

CHAPTER FIVE
- *Special Selections* 113

CHAPTER SIX
- *Combat Support Groups* 147

CHAPTER SEVEN
- *Desert Storm, Vietnam War,
and Current Trends* 165

Acknowledgements

First of all, may I express my sincere thanks and appreciation to all the contributors to these volumes of either historical knowledge or actual cadences. While I cannot begin to mention everyone involved with this project, there are some people and organizations whose special contributions helped make the compilation of these new editions and versions possible. I would like to acknowledge my family for their support, and especially my mother for her editorial assistance. My personal gratitude goes to my husband, Dr. Brian Johnson, simply for being there. Recognition is due to SGM William Lawrence (Fort Lewis) for his encouragement during the beginnings of the compilation. Acknowledgements are also due to *Army Times*, *Infantry Magazine*, *Leatherneck/Marine Corps Gazette*, the *Retired Officer Magazine*, *U.S. Army Aviation Digest*, and *Sergeants Magazine*.

Appreciation goes, too, to the following for use of cadences/information from both publications and individuals: *Victory Academy* (Drill Sergeant School), Fort Dix; 1st Aviation Brigade, *Jody Calls*, Fort Rucker; 2nd Armored Division, *Jody Calls*, Fort Hood; 104th Division Drill Sergeant School, Fort Lewis; U.S. Army Airborne School, Fort Benning; 82nd Airborne Division, Fort Bragg; 7th Transportation Group (Terminal), Fort

Eustis; *Jody Calls*, Fort Knox; 25th Infantry Division, Schofield Barracks; 4th Infantry Division (Mechanized), Fort Polk; Fort Sill; Fort Gordon; U.S. Army Band, Fort Meyer; Naval Training Center, San Diego; U.S. Air Force Academy; U.S. Naval Academy; Fort Lewis Military Museum; Fort Jackson Military Museum.

Certain individuals were particularly helpful: Gladys "Woodie" Borkowski, M/SGT, USAF (Ret.); J.E. Bennett (Curator of Military Music, Dept. of the Navy, HDQ. U.S. Marine Corps); John Slonaker (Chief, Historical Reference, Military History Institute, Carlisle Barracks); Jim Knight and Clay Blackburn (DAC Librarians); SFC G.L. Cook (Fort Carson); Major Bill Windsor, U.S. Marines, Ret.); Mrs. J.W. Varian (Lake Placid, NY); Midge Cornell (Fort Lewis); James Lutz (Toronto, Canada); Tim Frost (Dallas, TX); SFC Rick Dunlap (Fort McClellan); SFC Jim Gray (ROTC, Radford University); CPT Tom Burton (Fort Campbell); SGT Reynolds (U.S. Army Recruitment, Lynchburg, VA); staff and friends at Grandstaff Library, Fort Lewis.

Although I cannot list all their names here, gratitude is due to so many of the Public Affairs Officers all over the United States.

On a personal note, I'd like to thank special friends who listened and laughed with me: Mr. Kim TongHwi (Korea); Carmen "Tee" Thompson (Fort Gordon); Kyle Wilson (Camp Page, Korea); Dr. Abdullah Abdulkader (Saudi Arabia); Laurie Humphries (Fort Bragg); CW4 Kent Gallagher (Fort Lewis); Jack and Joyce Ullrich (White Sands

ACKNOWLEDGEMENTS

Missile Range); COL and Mrs. Philip Lauben (Clarksville, TN); Dr. Ronald Blazek (Florida State University); LTC Bob Vandel (MILPERCEN); LTC Mike Proctor (Fort Carson); CW2 Carla Baker (Fort Rucker); Marie Watson (Atlanta, GA).

Sandee Shaffer Johnson

* * * * * *

I will always thank my good friend Frank Sturgis, decorated WWII USMC veteran and a true American patriot, for his support and encouragement. He had courage enough for all of us.

Robert W. Morgan

Introduction

The new Jody Call handbooks from Talisman are not designed to be used as manuals jam-packed with technical details for use by instructors of precision drill. Instead, *The New Jody Call Handbook Series* by Talisman Books offers a wonderful compilation of rousing chants both old and very new for use by all proud members of every branch of the American Armed Forces.

The initial project was a direct result of numerous inquiries directed to Sandee Johnson's attention as a librarian for the U.S. Army. Many new recruits, NCOs, officers and even joggers were constantly searching for a handy, condensed book of provocative cadences. When she checked, she found to her surprise that no extensive collection had ever been published except by certain schools for Drill Sergeants, and an occasional enterprising individual with limited circulation and scope. By integrating such public domain material with that received from numerous contributors, she initially compiled two volumes for use among all the military branches as first published through Daring Books in 1983 and again in 1987.

Jody Calls are by their very nature wonderfully fluid because each generation of soldier likes to add their own experiences, desires and reflections to the rich base already provided by those who have marched before them. Therefore, at the urging of the editor-in-chief at Talisman Books, Robert W. Morgan, Sandee spent an entire year canvassing military bases the world over in search of the newest cadences generated by the Gulf War and the humanitarian experiences in Somalia.

Also, these two completely and newly revised and expanded editions are quite different in their scope, general content, and layout from her previous works. For instance, you will see that beside each *Jody Call* for the Air Force, Navy and Marines is affixed a service crest to indicate the originating branch of service. But that doesn't mean the Army can't use a *Jody Call* that originated in the Marine Corps it simply means these are the originators. Be imaginative and make 'em all your own!

While these two new volumes are a fairly complete cross-section of the cadences sung by the U.S.. military, it was impossible and counterproductive to include all the contributions that came across Sandee's desk. Had she done so, the resulting volumes would have been too thick to carry around, and that defeats our purpose. These volumes are meant to be used and enjoyed again and again.

INTRODUCTION

Some branches such as Infantry, Armor and Aviation related troops have contributed more chants than Support Groups. Unfortunately, there are a few lively Jodies which have only five or six words remaining after the vulgarities were edited; several more may warrant legitimate criticism from women due to their sexual undertones. However, Sandee's role as a compiler demanded that she remain impartial. Thus, she attempted to make these two volumes represent the best of what is currently being used by the American Armed Forces around the world, and that means the new as well as some the oldies-but-goodies.

Warning: For the critical grammarian, we might suggest they grit their teeth when they review these *Jodies*; they are rarely grammatically correct. However, *Jody Calls* always reflect current trends in American slang and specialized military jargon, and they seldom express love for the armed services. But they usually do reflect a positive tone, a wry sense of humor, and they contribute a certain atmosphere of spirit, pride, and a camaraderie seldom matched in civilian life. The role of *Jody Calls* in building "friendly competition" during physical training certainly cannot be underrated. In short, *Jody Calls* represent a microcosm that reflects the more popular thinking and emotions among the diverse patchwork of American citizens who have chosen to be trained to protect our country.

For those readers who possess an interest in their military heritage, these works include many of those wonderfully old and traditional marching songs that conform to the cadence beat as a drill leader sings out a line. In response, the troops either repeat his words, answer with a verse of their own, or count cadence. The common paces are *quick time* (marching) that refers to 120 steps (or beats) per minute while *double time* (running) indicates 180 steps (or beats) per minute.

Cadences have been used for centuries as a means to keep soldiers in rhythm or step. However, the *Jody Call* as we know it did not become widespread until the end of World War II. They are similar, however, to the Zouave chants that became popular during the American Civil War. Rumors of earlier military rhythmic chants still exist, but Sandee found no substantiation of these until after the publication of the *Duckworth Chant*.

Personally, Sandee recalled that as a child she sang some of the cadences without a clue as to their source. What could have been monotonous school bus trips and Girl Scout hikes were enlivened by versions of *Gee, Mom, I Wanna Go Home* or variations of *Captain Jack*. Many years later as a DAC in Korea, she was elated and thrilled when she recognized the spirited cadences shouted out by our troops as they double-timed down her street.

INTRODUCTION

Sandee has not knowingly infringed on any copyrighted material, and she gained whatever permission was necessary to use such material. But Jody chants seem to maintain the same distinction as folktales and dirty jokes—they change and grow from one generation to the next as they are verbally passed along. Some Jodies may be chanted to the tune of a popular song, or may even borrow phrases from well-known sources. In any case, Sandee has performed her due diligence to the best of her ability, and with good intent.

As time passes, some good old Jodies are irretrievably lost, and we find that sad. It is our fervent hope that these volumes, and our work in collecting and editing them, can serve as a preservation effort for still another facet of military lore. Just maybe, when the users of these volumes sing out loud and strong, some aging veteran or hardcore old drill sergeant who may be within earshot, will know their legacy is not totally forgotten.

Use them—dont lose them!

Sandee Shaffer Johnson, Compiler

Robert W. Morgan, Editor

I

This is Jody's story, sweet and short
Jody was born in a New York fort.
A man named Duckworth called "Sound off!"
Sound off - One, two, three, four,
One, two, Jo-dy ... Sound off!

A Brief History of Military Cadences

Cadences of different sorts have been utilized throughout the history of mankind as a means of maintaining a collective rhythm and coordinated pace while working, walking, jogging or running. For instance, on early galley ships in Roman and medieval times, a large drum was struck once for every stroke of the oars. Later, other musical instruments such as bugles and fifes were used to keep workers producing at a certain rate. Over the centuries, various working songs were developed in coal mines, railroads, and similar labor intensive industries to boost morale and promote esprit de corps among the workmen.

The Legions of the Roman Empire were among the first to discover that their massed infantry marched more effectively if they kept in step, and the innovative South African tribal war lord, *Shaka Zulu*, was quick to realize the benefits of having his warriors chant as they trotted mile after mile across the vast African *veldt* to wage war against their enemies. By the Eighteenth Century, defense systems among world powers had developed highly disciplined maneuvers, and most found efficiency in marching and drilling to the roll and beat of drums.

However, at the beginning of the American Revolution, the rag-tag volunteer Colonial Army was quite ignorant of the benefits of organized and regimented drill. Volunteer recruits who came both from cities and farms lacked even the most basic concept of military-style teamwork and organization that would promote effective battle maneuvers against the well-trained British forces; poorly trained American patriots paid dearly with their lives.

The story goes that the colonial farm boys were so ignorant, they literally did not know their left foot from their right. Keeping them in step was achieved by attaching a stalk of hay to one foot, and a piece of straw to the other so they could learn to march to America's first cadence call: *"Hay-foot, straw-foot, hay-foot!"* (This method was reportedly used again during the Civil War.)

Laughing at the scattered efforts of the "Rebel" troops as they tried to imitate an army, the Redcoats became so confident of their own superiority, that America's first and famous marching song *Yankee Doodle* was actually written by a British surgeon. Loudly sung with sneers and jeers by British troops, it was meant to be a cruel and disdainful taunt to all the inhabitants of the original Thirteen Colonies.

CHAPTER 1: A Brief History

Fortunately, Baron von Steuben, a professional Prussian Army officer engaged by the newly formed Continental Congress and serving under General Washington, quickly standardized military drills and uniform regulations among the green Yankee troops, thus contributing valuable discipline that led to the ultimate and glorious success of the first truly American Army. The war ended with American Independence, and the ragged but victorious former colonials lustily sang *Yankee Doodle* to every defeated British soldier they could catch.

Marching cadences quickly became an established part of military life in America, and the Civil War brought marching songs to the attention of the national conscience. Such melodies as *The Battle Hymn of the Republic* and *When Johnny Comes Marching Home Again* helped stir both the Union and the Confederate armies into battles filled with patriotic fury.

The first half of the twentieth centuries saw an influx of music, some dealing specifically with marching. *Tramp, Tramp, Tramp, Hip! Hip! Hooray! (We Are Marching Away)*, *Over There*, and the *Caisson Song* are only a few of the melodies that still haunt military Parade Grounds. Soon cadences themselves began to find mention in popular songs such as the official U.S. Army song, *The Army Goes Rolling Along* which boldly states *"...count off your cadence loud and strong!"*

The Legend of the Duckworth Chant

Toward the end of the Second World War, a new and revolutionary development in drill and marching surfaced among the American fighting personnel. As the story goes, in May, 1944, and as exhausted marching troops were dragging back to their barracks at Fort Slocum, New York, a unique and new rhythmic chant was suddenly heard from somewhere back in the columns. One black private by the name of Willie Duckworth wearily decided that the old dry *"Hup, two, three, four, Hup, hup!"* had lost its punch. He began singing out a short verse and then calling out *"Sound Off! One, two, three, four, one-two—three-four!"* Pretty quickly other soldiers began chiming in. Soon the chant spread throughout the columns and got louder and livelier with everyone picking up their dragging feet in sheer fun. Thus was born the famous *Duckworth Chant* or, as it is commonly known, *Sound Off!*

Luckily for every future G.I., Colonel Bernard Lentz, Fort Slocum's Commander at that time, instantly recognized the value of these unique verses. Having developed his own unique drilling methods designed to get troops motivated and sharp, he now ordered all his instructors to incorporate the *Duckworth Chant* from that day forward.

CHAPTER 1: A Brief History

Enter the War Department (now the Department of Defense). According to Gladys "Woodie" Borkowski, M/SGT, USAF (Retired) who happened to be stationed at Fort Slocum at the time, the *Duckworth Chant* was ordered to be recorded and distributed to the Armed Forces.

This was the result:

V Disc War Dept. Music Branch, Special Services Div. Army Service Forces.

Side A-1 Introduction Speech
Side A-2 T/Sgt Henry Felice and Rehabilitation

Side B-3 S/Sgt. Gladys Woodward & WAC Detachment, Fort Slocum, NY

  Jody Chant PVT James Tyrus & Rehabilitation Center Class

(Another source claims that Tony Martin, an extremely popular singer and actor from the 1940s, also recorded *Sound Off!* while he was stationed at Chanute Air Field, Illinois, during the Second World War.)

THE DUCKWORTH CHANT
Copyrighted © 1950 Bernard Lentz

The heads are up and the chests are out
The arms are swinging in cadence count.

CHORUS
(Leader)
> Sound Off!

(Troops)
> One, Two!

(Leader)
> Sound Off!

(Troops)
> Three, Four!

(Leader)
> Cadence Count!

(Troops)
> One, Two, Three, Four, One, Two
> ... Three-Four!

Head and eyes off the ground,
Forty inches, cover down.
(Repeat chorus)

It won't get by if it ain't G.I.
It won't get by if it ain't G.I.
(Repeat chorus)

I dont mind taking a hike,
If I can take along a bike.
(Repeat chorus)

I don't care if I get dirty,
As long as The Brow gets Gravel Gertie.
(Repeat chorus)

The Wacs and Waves will win the war,
So tell us what we're fightin' for.
(Repeat chorus)

They send us out in the middle of the night,
To shoot an azimith without a light.
(Repeat chorus)

There are lots of plums upon the tree,
For everyone exceptin me!
(Repeat chorus)

The first platoon, it is the best,
They always pass the Colonels tests.
(Repeat chorus)

Several other verses to the *Duckworth Chant* are usually grouped with the original stanzas:

(Leader)
    I (you) had a good home but I (you) left.
(Troops)
    You're right!
(Leader)
    I (you) had a good home but I (you) left
(Troops)
    You're right!
(Leader)
    Jody was there when you left.
(Troops)
    You're right!
(Leader)
    Jody was there when you left.
(Troops)
    You're right!
(Repeat chorus)

By this time everyone was getting in the act: The United States Coast Guard Band provided the sheet music to *Sound Off!* which added these lines:

The Captain rides in a jeep,
The Sergeant rides in a truck,
The General rides in a limosine,
But we're just out of luck!
(Repeat chorus)

In the preface to the Sixth Edition of Colonel Lentz's book *Cadence System of Teaching Close Order Drill and Exhibition Drills,* a portion of a letter attests to the popularity of the *Duckworth Chant* overseas. The correspondence is from Colonel Lentz's son, LTC Bernard V. Lentz, stationed at the time in Wurttemburg, Germany:

> *The Duckworth Chant (Fort Slocum edition) now resounds in the Third Battalion, 399th Infantry. The Companies are all pretty good at it by now. You should see the Dutchmen stick their eyes out when we go marching by with that number! The other day one of the Companies was taking a march and they started to chant. About five minutes after they moved out, they had more kids following them than the Pied Piper ever had; about five hundred from three years (old) up ... so enchanted were they by the Doughboys and their chant ...*
>
> Received shortly after VE Day

After World War II had wound down, cadence calls began to reflect various officers, units or posts, the introduction of new equipment, new drill instructors, and the constant rotation of troops. Many of the livelier marching songs were based upon black jive melodies, popular songs and sometimes even nursery rhymes. Even some good old spirituals such as *Amen, Amen,* or *"soul"* music were—and are—strutted to cadence throughout most U.S. military posts and bases.

As a study in evolving languages, cadences have a special niche all their own. Specialized vocabulary unique to wars and other contemporary circumstances are reflected in cadences for military personnel from all branches, just as poetry mirrors current events for students and civilians. For example, the Vietnam conflict, the Iranian crisis, and Desert Storm issue are frequent subjects of recent chants. Other topical trends such as streaking back in the '60's, rapid consumption of alcohol in the '70's, President Jimmy Carter's visit to Korea, the bombing of Muammar Khadafi's Libyan palace, and Saddam Hussein's desert challenge of the "mother of all battles" quickly found their way into a servicemen and servicewomens library of marching songs.

Who Is Jody, Anyway?

No one seems to know for certain when the *Duckworth Chant* (or *Sound Off!*) became known as the *Jody Call*, and subsequently was shortened to just *Jodies*.

For that matter, the common question over the years has been, *"Who is Jody?"* Your guess is as good as mine. But the consensus of opinions gathered from members of the military is not very complimentary! Soldiers of all ages and experiences agree that *"Jody"* is the rotten, lazy, sleazy, thieving, dope-smokin', draft-dodging, two-timing creep (guy or gal) back home ever—ready to take advantage of, or possibly swipe, your wife (husband), girlfriend (boyfriend), sister (brother), or even the family car. This Jody character is the career civilian who chooses to ignore the needs of his nation by enjoying the comforts of civilization while the sweating, serving, loyal and patriotic servicemen and servicewomen are risking their collective butts somewhere in the field or overseas.

The name *Jody* may have stemmed as an offshoot of "G.I. Joe", perhaps a variation of John Doe (J.D.), or maybe Willie Duckworth hated someone named Joe D. something-or-other. Whoever Jody is, legacy isn't too funny. If you try to make a joke about some good-livin' *Jody* to most G.I.'s during their "hardship" tours at some isolated post in Korea or Somalia, you won't get much laughter in response. While some of the old-timers might manage a rueful grin at the

CHAPTER 1: A Brief History

mention of Jody's name, the newer recruits may cringe at the implications.

Popular Jody Verses

What follows are some of the more popular verses about the ever-present *Jody* whose name pops up everywhere in conversations or jokes that don't really seem that funny to a homesick soldier. The chorus for these **Quick Time** cadences is usually:

Jody this and Jody that,
Jody is a real cool cat.

CHORUS
(Leader)
 Am I right or wrong?
(Troops)
 You're right!
(Leader)
 Am I going strong?
(Troops)
 You're right!
(Leader)
 Sound off!
(Troops)
 1, 2.
(Leader)
 Sound off!
(Troops)
 3, 4!
(Leader)
 Break it on down!
(ALL)
 1, 2, 3, 4, 1, 2 --- 3-4!

Ain't no use in calling home,
Jody's on your telephone.
(Repeat chorus)

Ain't no use in going home,
Jody's got your girl and gone.
(Repeat chorus)

Ain't no use in feeling blue,
Jody's got your sister, too.
(Repeat chorus)

Ain't no use in looking back,
Jody's got your Cadillac.
(Repeat chorus)

If old Jody's six feet tall,
I won't mess with him at all.
(Repeat chorus)

Might as well hide that frown,
Jody's beat you, hands down.
(Repeat chorus)

Jody, Jody, six feet four,
Jody's never been whipped before.
(Repeat chorus)

I'm gonna take a three day pass,
Can't wait to get Jody in my grasp.
(Repeat chorus)

Jody is the one who's mad,
Basic training ain't that bad!
(Repeat chorus)

A Soldier's Revenge

In the following cadence contributed by PFC James Knopp (Fort Hood, Texas), the Air Defense Artillery fights against Jody and wins.

A.D.A. versus Jody

Here I am in old Two-Two/Five,
Running PT—staying alive.
Back home Jody's got my wife,
I'll take a "Chap" and end his life.

CHORUS
(Leader)
>Sound Off!

(Troops)
>1, 2.

(Leader)
>Sound Off!

(Troops)
>3, 4!

(Leader)
>Break it on down!

(ALL)
>1, 2, 3, 4, 1, 2 --- 3-4!

Driving trucks the whole day through,
Watch out, Jody, I'm gunnin' for you.
So I went home, old Jody's scared,
Said, "Come on back, hey, if you dare."
(Repeat chorus)
Jody took an aircraft into the sky,
Said, hey, hey, Jody, you're gonna die.
Took that Chap and brought it on line,
Jody's death was mighty fine.
(Repeat chorus)

Jody's Got A Problem

History has not always been kind to the *Jody Call*. Some commanding officers have actually banned the use of cadences during drills, usually because of the obscenities and crude variations that pop up from time-to-time. Sure, they may be cute and fun at the moment, but they always get the *Jody* in trouble with the brass and visiting dignitaries. In more than one case, an officer or NCO has been relieved because his troops used vulgar cadences while running past high-ranking officers or buildings filled with of government workers.

Jody Goes To War

All that aside, *Jody's* have quickly carved their own special niche in our nation's proud military history. American warriors of all backgrounds, races, and creeds have tramped together to cadenced tunes as they trained to fight on the shores of Tripoli with the U.S. Marines, the great naval battles at Midway, the bloody race to the Rhine with the Army, and the spectacular low-flying raids over Baghdad with the Air Force. And as sure as *Jody* makes the recruit's life miserable, that same *Jody Call* will keep his or her stride quick, and sure, in step, and proud as hell!

We're the Jodies soldiers like best,
Mix us, match us — we'll do the rest,
So quick to learn 'n' easy to know,
We come in handy for those on the go.

Assorted Chants for Mixing and Creating

Chapter Two contains a variety of short chants and assorted lively chorus lines that challenge you to use them "as is", or to make up your own combinations. Make 'em different and lively!

CHANT VERSES

(To be mixed or matched with the choruses found at the end of this chapter)

I WANNA BE

I wanna be a Drill Instructor,
I wanna cut off all my hair.
I wanna be a Drill Instructor,
I wanna earn that Smokey Bear!

SAMPLE CHORUS
(Leader)
    Sound off!
(Troops)
    J! O!
(Leader)
    Sound Again
(Troops)
    D! Y!
(Leader)
    Bring it on down
(Troops)
    J.O.D.Y., J.O.—D.Y!

TARZAN & SUPERMAN

Jane winks her eye, old Tarzan swings,
Across the jungle and back again.
Lois flirts with Kent and Superman flies
Straight off a building one mile high.
(Insert selected chorus)

SERGEANT, SERGEANT

Sergeant, Sergeant, I have my doubts,
What's making your guts all stick out.
Is it whiskey or is it wine?
Or is it missing that P.T. time?
(Insert selected chorus)

THE MEANEST MAN

Standing in the sun all wet with sweat,
Drill Sergeant's the' meanest man I met.
Got a face like a monkey 'n' legs like a cat,
I didn't know anybody could look like that.
(Insert selected chorus)

SUPER NAVY

Spit shine those boots,
'N' shave off that hair.
Even though you can't compare,
Navy, we're Navy, super Navy!
(Insert selected chorus)

JANE FONDA

Jane Fonda, Jane Fonda, can't you see?
I gotta keep my country free.
Ain't no doper, ain't no way—
I'm the Number One hope of the U.S.A.
(Insert selected chorus)

BEATIN' THE DRUM

Sittin' on a mountain top, beatin' on a drum—
Beat so hard that the M.P.s come.
M.P., M.P., don't get me—
Get that Leg behind that tree.
(Insert selected chorus)

EARLY ATTACK

Up in the morning, outta that sack,
Greeted at dawn by an early attack.
First Sergeant rushes me off to chow,
But I don't eat it anyhow.
(Insert selected chorus)

BIRDIE, BIRDIE

Birdie, Birdie, in the sky,
Dropped a little white wash in my eye.
Ain't no sissy, I won't cry—
I'm just glad that cows don't fly.
(Insert selected chorus)

ICE COLD BEER

What the hell we doin' here?
Let's go out and get a beer.
Ice cold beer with lots of foam,
Like the kind we get at home.
(Insert selected chorus)

SAINT PETER

Heaven is great, to my surprise,
There's a lot of Airborne guys.
There stands Saint Peter on the crest,
He has wings upon his chest.
(Insert selected chorus)

BODY BAGS

When the day is done,
And taps is played,
They'll lower our flag,
Over body bags.
or
When the day is done,
And taps is played,
They'll lower their flag,
Over their body bags -
Don't mess with us,
We're U.S.!
(Insert selected chorus)

KILLER P.T.

Drill Sergeant, Drill Sergeant, can't you see,
This P.T. is killing me!
I've got pain in all my chest,
I might die, but I'll do my best!
(Insert selected chorus)

GIRL FROM NEW ORLEANS

I had a girl in New Orleans,
Fourteen kids and a can of beans.
Now she's someone else's wife,
And I'll be runnin' for the rest of my life.
(Insert selected chorus)

PISS IN THE BOTTLE

I piss in the bottle and come up hot,
Now the First Sergeant's on my back.
He recommended me to D and A,
Now they march us there every day.
(Insert selected chorus)

BAGHDAD BETTY

Baghdad Betty's crying,
Cause her troops are dying.
Come on Betty, dry your eyes,
We'll stuff your mouth with apple pies.
(Insert selected chorus)

'ROUND THE WORLD

Airborne, Airborne, where've you been?
'Round the world and goin' again.
What'll you do when you get back?
Gonna get me a new rucksack!
(Insert selected chorus)

CO. 378

Get off the grinder 'cause here's a reminder,
That 3-7-8 is sure enough finer.
And if you don't, you'll wish you had,
'Cause 3-7-8 is sure enough bad.
(Insert selected chorus)

THE GIRL OUT WEST

I had a girl who lived out west,
Thought this Airborne life was best.
Now she's somebody else's wife,
I'll be jumpin' for the rest of my life.
(Insert selected chorus)

LITTLE BITTY FEET

Listen to the rhythm of the little bitty feet,
Sounds like the Army in full retreat.
In the Army, young and old,
They wanna wear the red and gold.
(Insert selected chorus)

TRANSPORTATION BLUES

Daddy, Daddy, look and see,
What Transportation's done to me.
Driving jeeps and flyin' planes,
Now I'll never be the same.
(Insert selected chorus)

THE BOLD INFANTRY

Out in the dark, ready to fight,
U.S. Infantry has the might.
In the snow and in the cold,
U.S. Infantry is mighty bold.
(Insert selected chorus)

AHAB'S CAMEL

Ahab had a camel named Clyde,
Nobody knows how the poor boy died.
He rode all day and he rode all night,
My boy Ahab rode out of sight.
(Insert selected chorus)

STRAIN FOR SEMPER FI!

If you want it,
you've got to earn it.
It ain't easy!
You've got to work for it,
You've got to strain for it!
Marine Corps, Semper Fi!
(Insert selected chorus)

BOW-LEGGED WOMAN

There's two things that I can't stand,
A bow-legged woman, and a straight-legged man.
(Insert selected chorus)

COMBAT TRIP

Huey, Cobra, sittin' on the strip,
Army aviator on a combat trip.
Jump in, buckle up, close the canopy,
Never mind the rest, keep your eyes on me.
(Insert selected chorus)

CHAPTER 2: Assorted Chants

NO-MAN'S LAND

If I can't get through that Heaven's gate,
I'm going to Hell to suffer my fate.
Grease gun and K-bar in my hand,
Turn that place into no-man's land.
(Insert selected chorus)

BLISTERS AND A TAN

Hittin' the track 'fore coffee's hot,
Wish like hell I was back in my cot.
Dad said Basic would make me a man,
But all I'm getting is blisters and a tan.
(Insert selected chorus)

LOOSE LIPS

Loose lips sink ships,
Then the Navy's got to save 'em.
(Insert selected chorus)

AWOL IN WAHIAWA

AWOL, AWOL, where you been?
Down in Wahiawa drinkin' gin.
What'cha gonna do when you get back?
Sweat it out on the P.T. track.
(Insert selected chorus)

STICK WITH IT!

One, two, three, four,

If we don't sing, we're gonna run some more.

One mile—

Won't get it!

Two miles—

Stick with it!

Three miles—

Lookin' good!

Four miles—

Knew we could!
(Insert selected chorus)

HARD DAY'S WORK

A hard day's work is all we know,
That's the Infantry wherever we go.
Those MP's know we do raise hell,
But in a war, we'll do well!
(Insert selected chorus)

MEAN OLD INFANTRY

MPs know we raise some hell,
But in the battle, why we're so fine.
So if you ask why we're so fine,
We're the mean old Infantry line!
(Insert selected chorus)

PEARLY GATES

One, Two, three, four,
God bless the Marine Corps,
'N' open up the Pearly Gates.
Nine, ten, eleven, twelve,
All the rest can go to hell.
(Insert selected chorus)

MY OLD GRANNY

My old Granny is seventy-two,
Knows karate and a little Kung Fu.
Bustin' boards and breakin' bricks,
Knocks big trees into pick-up sticks.
(Insert selected chorus)

REDLEG SOLDIER

Redleg soldier, implace your spade,
Before the battery gets laid.
Set the collimater in time,
Before the enemy gets behind.
(Insert selected chorus)

RUBBER BAND BLUES

Up every mornin' at a quarter t'six,
Jumpin' and runnin' and gettin' sick.
Bending and stretchin' like a rubber band,
Lord, oh Lord, won't you give me a hand?
(Insert selected chorus)

TURNIN' MEAN

Marine Corps put me in a barber's chair,
And shaved off all my civilian hair.
Put me in these cammy greens,
Made me run 'til I'm turnin' mean.
(Insert selected chorus)

INFANTRY GUTS

We'll hit your lines and split you in two,
You'll be runnin' yellow before we're through.
We'll fight with every weapon in our hand,
Because that's the guts of an Infantry man.
(Insert selected chorus)

HITCHHIKER BLUES

Standing on the highway a-thumbing a ride,
A bus load of paratroopers pass me by.
Hanging out of the window was a
bald-headed man,
He yelled some things I couldn't understand.
(Insert selected chorus)

CHAPTER 2: Assorted Chants

GEE, I ...

G.I. coat and G.I. comb,
Gee, I wish that I was home.
G.I. coat and G.I. gravy,
Gee, I wish I'd joined the Navy!
(Insert selected chorus)

MERCY ME!

Oh, mercy yes, oh, mercy me,
There's a new breed of men in Bill's army.
Wish I may, wish I might,
That _____ are not allowed to fight!
(Insert selected chorus)

PRIVATE JONES

You're a grunt now, Private Jones,
No private room or telephones.
You had meals in bed before,
You won't have 'em there anymore.
Mr. Jones wore faded jeans,
Private Jones wears O.D. greens!
(Insert selected chorus)

ACHING BACK BLUES

Every night I hit the sack,
Oh, my aching Airborne back!
(Insert selected chorus)

FOUR MILES OUT

When it's over let me do it again,
Four miles out and four miles in.
Gunpowder for breakfast and J.P.4,
Run like hell and ask for more.
(Insert selected chorus)

RIDE A BIKE

We like running, yes, we do!
And if you tried it, you would, too.
One thing better we would like,
We would rather ride a bike.
(Insert selected chorus)

TRANSPORTATION PRIDE

Hey, hey, Transportation,
I've got my eyes on you.
You know you look so straight and neat,
When you're marching to the beat.
(Insert selected chorus)

ARMOR CREWMAN

I want to be an Armor crewman,
Live a life that's almost human.
In my tank I feel no dangers,
I run all over Airborne Rangers.
(Insert selected chorus)

ROCK 'N' ROLL

Don't you know, Rock 'n' Roll,
In my heart, in my soul.
Gotta go, all the way!
Rock 'n' Roll is here to stay.
(Insert selected chorus)

THE GIRL IN BROWN

I know a girl all dressed in brown,
Makes her living goin' up and down.
Elevator operator, Deep sea diver!
(Insert selected chorus)

LET ME DRIVE

If I live forever more,
Don't send me to the Five-Oh-Four.
Send me down where I can drive,
Way on down to the Five-Five-Five.
(Insert selected chorus)

EYES UP!

Keep your eyes up off the ground,
Ain't no discharge layin' round.
Don't you stop, regardless of rank,
The man behind you is drivin' a tank!
(Insert selected chorus)

STRAIGHT-LEG BATTLE CRY

The battle cry of the straight-leg corps,
"Medic, Medic, I'm so sore!"
The battle cry of the straight leg corps,
"Had some leave and want some more!"
(Insert selected chorus)

GOOD AS GOLD

I don't know but I've been told,
The _____ is good as gold.
(Insert selected chorus)

AIRBORNE RANGER

Airborne Ranger, ravin' mad,
He's got the patch that I wish I had.
Yellow and black and a half-moon shape,
Airborne Ranger, he's gone ape!
(Insert selected chorus)

FUN IN YOUR RUN

Looky here comin' into sight,
Mornin', girls, you're looking nice.
Won't you come and run with me?
We're the _____ Artillery.
We will run 'til the settin' sun,
We'll put more fun in your run.
(Insert selected chorus)

GOIN' OVERSEAS

I don't know, but I believe,
You'll be goin' overseas!
(Insert selected chorus)

MILITARY LEFT

Your left, your left,
Your left, right, left!
Your military left.
Your left, your right, now pick up the step.
Your left, your right, your left!
(Insert selected chorus)

GOIN' STRONG

(Leader)
    Am I right or wrong?
(Troops)
    You're right!
(Leader)
    Am I right or wrong?
    ... or Ain't we goin' strong!
(Troops)
    You're right!
(Insert selected chorus)

NOW FOR THE CHORUSES

To be Inserted with your favorite verse

The following chorus chants and responses are a collection from all around the military bases. They are meant to be mixed and matched with the preceeding **Verses** in any way you wish. Be creative and shout 'em out.

SOUND OFF JODY

(Leader)
   Sound off!
(Troops)
   J.O.
(Leader)
   Sound Again
(Troops)
   D.Y.
(Leader)
   Bring it on down -
(Troops)
   J.O.D.Y., J.O.—D.Y!

HIDI-HIDI-HO!

(Leader)
  Hidi, Hidi, Hidi, Ho,
(Troops)
  Hidi, Hidi, Hidi, Ho!
(Leader)
  Hidi, Hidi, Hidi, Hey,
(Troops)
  Hidi, Hidi, Hidi, Hey!
(Leader)
  Hidi, Hidi, Hidi, Hey,
(Troops)
  Hidi, Hidi, Hidi, Hey!
(Leader)
  Hidi, Hidi, Hidi, Hey,
(Troops)
  Hidi, Hidi, Hidi, Hey!
(Leader)
  Hidi, Hidi, Hidi, Hey,
(Troops)
  Hidi, Hidi, Hidi, Hey!
(Leader)
  Hidi, Hidi, Hidi, Ho,
(Troops)
  Hidi, Hidi, Hidi, Ho!

GET ON DOWN

(Leader)
   Left right, get on down.
(Troops)
   Left, right, get on down!
(Leader)
   Left right, get on down.
(Troops)
   Left, right, get on down!
(Leader)
   Get on down,
(Troops)
   Get on down!
(Leader)
   Get on down,
(Troops)
   Get on down!
(Leader)
   Get on down now,
(Troops)
   Get on down now, left, right,
   ... get on down! WHOA, WHOA!
(Leader)
   Whoa, whoa, whoa, who!
(Troops)
   Whoa, whoa, whoa, who!
(Leader)
   Whoa, whoa —— whoa-whoa!
(Troops)
   Whoa, whoa —— whoa-whoa!

OLLY ANNA!

(Leader)
>Olly Anna, Olly, Olly Anna,

(Troops)
>Olly Anna, Olly, Olly Anna!

(Leader)
>Olly, Olly, Olly, Olly, Olly, Olly Anna.

(Troops)
>Olly, Olly, Olly, Olly, Olly, Olly Anna!

(Leader)
>Dress it right and cover down,
>... Olly, Olly Anna,

(Troops)
>Dress it right and cover down,
>... Olly, Olly Anna!

(Leader)
>Forty inches all around, Olly, Olly Anna,

(Troops)
>Forty inches all around, Olly, Olly Anna!

(Leader)
>Olly Anna, Olly, Olly Anna,

(Troops)
>Olly Anna, Olly, Olly Anna!

(Leader)
>Olly, Olly, Olly, Olly, Olly, Olly Anna,

(Troops)
>Olly, Olly, Olly, Olly, Olly, Olly Anna!

III

We are chanted whenever you drill,
Sing us in fields, in valleys or hills.
Though many of these Jodies are much older,
Just like Vets, they're sometimes bolder!

General Cadences

DOUBLE TIME

Double time, double time up the hill,
Everybody's gonna get a two mile thrill.
Double time, double time, everyone will,
Everybody's gonna get their fill!

Double time, double time, two miles long,
How in hell can we go wrong?
Double time, double time to this song,
Everybody's gonna make it strong!

Double time, double time, going strong,
Who ever thought we could run so long?
Double double time all the way,
We get up and run all day!

Double time, double time, love it, too,
We can run the whole day through.
Double time, double time, everybody's havin' fun,
We can't wait for the four mile run!

MISTER P.T.

Mister P.T. is playing my song,
Running fast, marching long.

Mister P.T. is singing my note,
Cut it short, you'll get my vote.

Mister P.T. is tooting my tune,
Hot the shower, hit it soon.

Mister P.T. is attempting to shine,
Marching all the way up the Rhine.

Mister P.T. is learning to rap,
Gonna dance Karlsruhe on the map.

Mister P.T. is trying to roll,
He's even trying to run to soul.

Mister P.T. is making me see double,
We'd better quit before he's in trouble!

Anonymous, Karlsruhe, Germany

BURGER KING

Down in Honolulu at the Burger King,
First Sergeant _____ was a-doin' his thing.

Hamburger, hot dog, chocolate shake,
There isn't too much that he can't take.
Stand up, hustle up, shuffle to the door,
Back to the track and run some more.

HQ, 25th Infantry Division, Schofield Barracks, HI

CAPTAIN JACK

Hey, hey, Captain Jack,
Meet me down by the railroad track.
With my rifle in my hand,
I'm gonna be a fightin' man.

Hey, hey, Captain Jack,
Meet me down by the railroad track.
With my suitcase in my hand,
I'm gonna be a traveling' man.

Hey, hey, Captain Jack,
Meet me down by the railroad track.
With my car keys in my hand,
I'm gonna be a drivin' man.

Hey, hey, Captain Jack,
Meet me down by the railroad track.
With my bottle in my hand,
I'm gonna be a drinkin' man.

Inoy Yague, Fort Lewis, WA

IN SHAPE

Round the post and round we go,
Where we'll stop only top knows.

Just for fun we hump thirty pound packs,
Bouncing real hard, bruising our backs.
For the stars and stripes we'll run,
For Old Glory we'll fire our guns.

Anonymous, Fort Lewis, WA

KNUCKLEHEAD

Oh, you knucklehead,
Dumb, dumb knucklehead!
Marchin' down the avenue,
___ more weeks and we'll be through!

Oh, you knucklehead,
Dumb, dumb knucklehead!
I'll be glad and so will you,
When all this trainin' here is through.

Oh, you knucklehead,
Dumb, dumb knucklehead!
Goin' to A.I.T.
There's more in store for you and me!

Army Recruitment Office, Lynchburg, VA

BETTER STEP ASIDE

If you see me coming,
Better step aside.
A lot of men didn't -
A lot of men died.

One fist of iron,
The other of steel.
If the right one don't get you,
Then the left one will!

SPC C. Williamson, Jr., 3/37th FA, Erlangen, Germany

A LETTER FROM MY RECRUITER

I was sitting at home watchin' TV,
Drinking beer at a quarter to three.
Up walked the postman and dropped the bomb,
He handed me a letter from the Pentagon.

My knees got shaky and I began to sweat,
I said "I know they haven't started the draft
up yet."
I opened up the letter and what did I see?
A whole lot of big words that I couldn't read.

This Army is supposed to be good for me,
And so I went downtown to see.
The Army Recruiter said, "Don't despair,
Opportunity unlimited, and treatment that's fair."

Pay is good, advancement is great,
Get out of bed early, go to bed late.
I mean what I say, and say what I mean,
Army life is the best I've seen.

Victory Academy, U.S. Army Drill Sergeant School, Fort Jackson, SC

I'M NOT ...

I don't know, but I think I might,
Jump from an airplane, while in flight.
Got three kids, gonna have three more,
Two on the ground and one in the door.
I'm not the preacher, or the preacher's son,
But I'll take the money 'til the preacher comes.
I'm not the butcher or the butcher's son,
But I'll do the killing 'til the butcher comes.
I'm not the mama or the mama's son,
But I'll do the lovin', 'til the mama's son comes.

HQ, 25th Inf. Div., Schofield Barracks, HI

HEY, HEY, GOODBYE!

Sha Na Na Na, Na Na Na Na,
Hey, hey, hey, goodbye.

No more roadmarks, no more doing sharks,
Hey, hey, hey, goodbye.

No more M.E.Ds. hiding in the trees,
Hey, hey, hey, goodbye.

No more P.T., P.T. hurts my body,
Hey, hey, hey, goodbye.

No more E.P.O., 'cause he doesn't know,
Hey, hey, hey, goodbye.

No more A.P.O., 'cause he runs too slow,
Hey, hey, hey, goodbye!

Co. 351, Naval Training Center, Great Lakes, IL

HEY, BOBBA REEBA (Marching)

Well, I wish all the ladies,
Were bricks in a pile.
And I was a mason—
I'd lay 'em all in style.

CHORUS
Hey, Bobba Reeba,
Hey, Bobba Reeba,
One-two ---- three, four!
One-two ---- three, four!

Well, I wish all the ladies,
Were pies on a shelf.
And I was a baker—
I'd eat 'em all myself.
(Repeat chorus)

Well, I wish all the ladies,
Were fish in the sea.
And I was a kingfish—
I'd have 'em all for me.
(Repeat chorus)

Well, I wish all the ladies,
Were clouds in the sky.
And I was a pilot—
They'd all help me fly!
(Repeat chorus)

104th Division, Drill Sergeant School, Fort Lewis, WA

I WANNA GO HOME

The meat in the Army,
They say is mighty fine.
Last night we had ten puppies,
Today we've only nine.

CHORUS
I don't want no more Army life,
Gee, Mama, I wanna go,
But they won't let me go,
Gee, Mama, I wanna go home!

The coffee in the Army,
They say is mighty fine.
It's good for cuts and bruises,
And tastes like iodine.
(Repeat chorus)

The stockings in the Army,
They say are mighty fine.
But they're not very sheer 'cause,
They're made of binder twine.
(Repeat chorus)

The shoes in the Army,
They say are might fine.
You ask for number sevens.
They give you number nines.
(Repeat chorus)

Romances in the Army,
They say are mighty fine.
But just like ripe tomatoes,
We're rotting on the vine.
(Repeat chorus)

The pancakes in the Army,
They say are mighty fine.
But when you try to chew them,
You only waste your time.
(Repeat chorus)

M. Cornell, Fort Lewis, WA

THEY SAY THAT IN THE ARMY

They say that in the Army, the chicken's mighty fine,
One jumped off the table, and started marking time.

CHORUS
Oh, Lord, I want to go,
But they won't let me go,
Oh, Lord, I want to go,
Home, home, home...

They say that in the Army, the pay is mighty fine,
They pay you a hundred dollars, and take back 99.
(Repeat chorus)

They say that in the Army, the coffee's mighty fine,
It looks like muddy water, and tastes like turpentine.
(Repeat chorus)

They say that in the Army, the women are mighty fine,
They look like Phyllis Diller, and walk like Frankenstein.
(Repeat chorus)

They say that in the Army, the biscuits are mighty fine,
But one fell off the table, and killed a friend of mine.
(Repeat chorus)

Anonymous

Note: In the preceding cadence, They Say That In The Army, many verses may be substituted. In the same spirit, I Wanna Go Home has cadences utilized by the WACs during earlier times.

ARMY BOOTS ARE MADE FOR WALKIN'

Note: This is one of the cadences that was adapted from These Boots Were Made For Walking as recorded by Nancy Sinatra.

These boots are made for walkin',
And that's just what they'll do,
If all you're doin' is markin' time,
They'll walk all over you.

These guns are made for shootin',
And that's just what they'll do.
And if we get a mission,
We'll drill a hole in you.

This Army's trained for fightin',
And that's just what we'll do.
And if you pick a fight with us,
We'll run all over you.

Victory Academy, U.S. Army Drill Sergeant School, Fort Jackson, SC

SHAKE, RATTLE AND ROLL

Shake, rattle and roll, everybody!
Shake, rattle and roll, everybody!
Boom chakalaka, ah-ah, ah-ah,
Boom chakalaka, ah-ah, ah-ah!
I'm gonna run all day!

Boom chakalaka, ah-ah, ah-ah!
Party all the live-long night!
Boom chakalaka, ah-ah, ah-ah!
Run real hard 'til the broad daylight!

Co. C, 16th Ord. Bn, 61st Ord Bde, Aberdeen Proving Ground, MD

EENY-MEENY-MINY-MO

Shoot those guns, fire 'em high or low,
Eeny-Meeny-Miny-Mo.

We're the ones they call G.I. Joe,
Eeny-Meeny-Miny-Mo.

Catch that Commie by his toes,
Eeny-Meeny-Miny-Mo

His the target and kill the foe,
Eeny-Meeny-Miny-Mo.

Anonymous, Fort Eustis, VA

PT

Your eyes are right,
Your pants are tight.
Your packs are swinging,
From left to right.

CHORUS
Use Sound Off! chorus.

Do not worry,
Do not fret.
Tell your sergeant,
and let him sweat.
(Repeat chorus)

Wm. Cave, Summerland Key, FL

RUNNING ALL THE WAY TO ...

Up in the mornin' 'bout a quarter t' three,
My First Sergeant was bringin' heat.
Had NCOs all around his desk,
And a full-bird Colonel in the leanin' rest.

First Sergeant, First Sergeant, can't you see?
You can't bring no smoke on me.
I can run to Denver running like this,
All the way to "D" town running like this.

How about L.A. running like this,
All the way to California running like this.
I can run to Kansas City running like this,
All the way to Chief's Town running like this.

SFC G.L. Cook, EO, HQ, 3rd Bde, 4th Inf. Div., Fort Carson, CO

CHAPTER 3: General Cadences

Note: The following two marching chants (Chesty Puller and Old King Cole) exemplify the natural variations that develop when cadences are passed by word of mouth and the competition and rivalry between members of the U.S. Marine Corps and the U.S. Army. Both chants are adaptations of the old nursery rhyme "Old King Cole." For each call, the chorus is repeated by the troops and, as the song progresses, different answers are shouted out by changing ranks.

CHESTY PULLER

Chesty Puller was a good Marine,
And a good Marine was he!
He called for his knife,
And he called for his gun,
And he called for his Privates Three.

Booze, booze, booze, said the Privates.

CHORUS
Merry men are we!
There's none so fair,
That can compare,
With Marine Corps Infantry.

We don't give a damn, said the Corporals,
Booze, booze, booze, said the Privates.
(Repeat chorus)

Get that squad in step, said the Sergeants,
We don't give a damn, said the Corporals,
Booze, booze, booze, said the Privates.
(Repeat chorus)

(high-pitched voice)
We do all the work, said the Louies,
Get that squad in step, said the Sergeants,

We don't give a damn, said the Corporals,
Booze, booze, booze said the Privates.
(Repeat chorus)

Shine my shoes and brass, said the Captains,
We do all the work, said the Louies,
Get that squad in step, said the Sergeants,
We don't give a damn, said the Corporals,
Booze, booze, booze said the Privates.
(Repeat chorus)

How 'bout junk on the bunk, said the Majors
Shine my boots and brass, said the Captains,
We do all the work, said the Louies,
Get that squad in step, said the Sergeants,
We don't give a damn, said the Corporals,
Booze, booze, booze said the Privates.
(Repeat chorus)

How 'bout a three day pass, said the Colonels,
How 'bout junk on the bunk, said the Majors,
Shine my boots and brass, said the Captains,
We do all the work, said the Louies,
Get that squad in step, said the Sergeants,
We don't give a damn, said the Corporals,
Booze, booze, booze said the Privates.
(Repeat chorus)

War, war, war, said the Generals,
How 'bout a three day pass, said the Majors,
Shine my boots and brass, said the Captains,
We do all the work, said the Louies,
Get that squad in step, said the Sergeants,
We don't give a damn, said the Corporals,
Booze, booze, booze said the Privates.
(Repeat chorus)

MORE VERSES!

There are other Marine versions to this particular cadence that can be varied by inserting one or more of the following.

Of course, you may have you own words ... ?

Chesty Puller was a grand old man,
And a grand old man was he.
He called for his pipe,
And he called for his bow,
And he called for his Generals Three.

What'cha gonna do when I am gone, said Chesty.
We don't have a clue, said the Generals.
Can't you lead your men? said the Colonels.
You'll never get that leave, said the Majors.
Kiss that fit rep goodbye, said the Captains.
What the hell is that, said the Gunnies.
Get your men in step, said the Sergeants.
Left, right, left, said the Corporals.
Beer, beer, beer! said the Privates.

SGT Clements, USMC, Camp Pendleton, CA

OLD KING COLE

This is the U.S. Army version and is chanted the same as Chesty Puller.

Old King Cole was a merry old soul,
And a merry old soul was he,
He called for his pipe,
And he called for his bowl,

And he called for his Privates Three.

Beer, beer, beer, said the Privates.

CHORUS
Fighting men are we!
But there's none so fair that they can compare,
with the Airborne Infantry (or insert name of unit)

Repeat as in "Chesty Puller" chant by adding the following:

Corporals Three: Where's my pass, said the Corporals.

Sergeants Three: Work, work, work, said the Sergeants.

Louies Three: What'll we do now? said the Louies.

Captains Three: Take that hill, said the Captains.

Majors Three: Cover my tracks (a__?), said the Majors.

Colonels Three: Fight, fight, fight, said the Colonels.

Generals Three: Win, win, win, said the Generals.

HOA, WHOA, WHOA, WHOA

I used to love a high school queen,
Now I love an M-16.

CHORUS
Whoa, whoa, whoa, whoa,
Whoa, whoa ----- whoa-whoa!

I used to drive a Chevrolet,
Now I'm running every day.
(Repeat chorus)

I used to love my high school queen,
Now I'm wearing this Army green.
(Repeat chorus)

SGT R. Smith, Fort Lewis, WA

VOICE OF THUNDER

I want to be a Drill Instructor,
I want to cut off all my hair.

I want to be a Drill Instructor,
I want to earn that Smokey Bear!

***USMC Drill Instructor Chant Book**, Marine Corps Recruit Depot, San Diego, CA*

COUNT CADENCE

(Leader)
　　Count cadence, delayed cadence,
　　Count cadence, count!
(Troops)
　　One!
(Leader)
　　I'm a soldier,
(Troops)
　　Two!
(Leader)
　　And I do my best.
(Troops)
　　Three!
(Leader)
　　But each day I find myself,
(Troops)
　　Four!
(Leader)
　　In front of the C.O.s desk.
(Troops)
　　One!.
(Leaders)
　　Hit it!
(Troops)
　　Two!
(Leader)
　　Hit it!
(Troops)
　　Three!
(Leader)
　　Hit it!
(Troops)
　　Four!

CHAPTER 3: General Cadences

(Leader)
    Get down!
(ALL)
    One, two, three, four,
    One, two — three, four!

(Repeat in same Leader/Troops format as previous)

Count cadence, delayed cadence,
Count cadence, count!
One, heads up,
Two, shoulders back,
Three, don't be looking down,
Four, get your feet off that ground!
One, hit it!
Two, hit it!
Three, hit it!
Four, get down!
One, two, three, four,)
One, two, three, four!

SFC K. G. Dailidonis, 800th CMMC, Germany

DRIP, DRIP, DRIPPITY DROP, DROP!

(Leader)
Rain is falling, water dripping on my head.
(Troops)
Drip, drip, drippity drop, drop!

(Leader)
Roof is leaking, water dripping in my bed.
(Troops)
Drip, drip, drippity, drop, drop!

(Leader)
Can't go out so I might as well be dead.
(Troops)
Drip, drip, drippity, drop, drop!

(Leader)
PT, PT, PT, PT everyday.
(Troops)
Drip, drip, drippity, drop, drop!

(Leader)
PT, PT, PT, PT the Marine Corps way.
(Troops)
Drip, drip, drippity, drop, drop!

(Leader)
Love to build my body the Marine Corps Way.
(Troops)
Drip, drip, drippity, drop, drop!

SSG K. M. Bemis, USMC, Camp Pendleton, CA

ARMY LIFE

Oh, I joined the Army ranks,
Just so I could see a tank.

Oh, I left a nagging wife,
Just to lead this Army life.

Oh, this Army is for me,
It's the only place to be.

I like women and I like wine,
But all I do is double time.

Double time here and double time there,
Man, this life, it's the best anywhere.

Victory Academy, U.S. Army Drill Sergeant School, Fort Jackson, SC

CAPTAIN CARR

Hey, hey, Captain Carr,
Meet me down at the nearest bar.
With my git-tar in my hand,
I wanna be a singin' man!

Hey, hey, Captain Carr,
Meet me down at the nearest bar.
With my video in hand, I wanna to be a VCR man.

Hey, hey, Captain Carr,
Meet me down at the nearest bar.
With a boom-box in my hand,
I wanna be a dancin' man.
Anonymous, Fort McClellan, AL

ALL THE WAY

Hey, hey, all the way!
We love to run every day.

If I was the President and had my way,
There wouldn't be a fat man in the Army today.

Everyone would be fit to fight,
Whether you test 'em day or night.

When I jumped onto that old drop zone,
Most of the enemy had already gone.

Those who remained weren't fit to fight,
So enemy contact was really light.

We ran the stragglers off the old drop zone,
Everything is quiet and they're all gone.

I ran towards an improved machine gun nest,
Spraying lead, I was really at my best.

The enemy tried to bob and weave,
My blood-curdling screams like to make him heave.

I snatched him out of his well-dug hole,
And really fixed him, God bless his soul!

CSM F. Gerber, 1st Inf. Div. NCO Acad., Fort Riley, KS

NAVY BLUES

Hup! 2, 3, 4.
S and S, then out the door.

Hup! 2, 3, 4.
Hit the decks, mop that floor.

Hup! 2, 3, 4.
Jog twice around, then twice more.

Hup! 2, 3, 4.
Dad was right—it's a bore.

Hup! 2, 3, 4.
All I'm hoping is a night on shore.

Hup! 2, 3, 4.
Off the coast just waiting for war.

Hup! 2, 3, 4.
Wishing for the mail a little more.

Hup! 2, 3, 4.
So this is called a six-month tour.

Naval Training Center, Great Lakes, IL

HELLO, JOSEPHINE

Hello, Josephine, how do you do?
Do you remember me, baby, like I remember you?

You used to live over yonder, by the railroad tracks,
And every time it rained, you used to call my name.

We used to meet by the lake, you'd make my poor heart ache,
And by that old moonlight, you'd always hold me tight.

Used to meet beside the stream, and how we'd always dream,
And when we'd meet at night, you'd always treat me right,

And when the moon would shone, we used to wine and dine.
Hello, Josephine, how do you do?

Do you remember me, baby, like I remember you?

Drill Sergeants School, Fort Dix, NJ

VITAMIN P

Old John Wayne was a friend of mine,
We did P.T. all the time.

Push up, sit up, two-mile run,
We didn't stop 'til all was done.

John Wayne loved his Vitamin P,
He taught me what's good for me.

On Day One, I was puny and weak,
John Wayne started with the bend-and-reach.

On Day Two, it was doin' me good,
I kept my faith, like he knew I would.

On Day Three, I was tall and proud,
I felt so good, I led the crowd.

P.T.s good for you and me,
We'll never O.D. on Vitamin P.

We do P.T. the John Wayne way,
We do P.T. every day.

I like P.T. and that's no lie,
I'll do P.T. 'til I die.

P.T. keeps me fit and strong,
With Vitamin P - P.T! Vitamin P - P.T!

SFC N. W. Fox, Btry. B., 1st Bn. 230th FA, Reidsville, GA

CALIFORNIA DREAMIN'

In Monterey the days are long,
That's why there's time to sing this song.
We march all day and dance all night,
This California livin' is out of sight.

Tourists come here to see the coast,
For us it's just another post.
We must say but won't repeat,
hose California gals sure are neat!

Anonymous, Fort Ord, CA

B-A-R

(To the tune of the Crawdad Song)

I don't want a B-A-R, honey, honey!
I don't want a B-A-R, babe, babe!
I don't want a B-A-R—
All I want is a steel guitar, honey, oh baby, mine!

I don't want to march no more, honey, honey!
I don't want to march no more, babe, babe!
I don't want to march no more—
Put me in the Signal Corps!

Anonymous, PAO, Fort Polk, LA

SOLEMN VOW

Through the desert, across the plains,
Steaming jungle and tropic rains.
No mortal foe can stop me now—
This is gonna be my solemn vow.

I have honor, and I have pride—
Winning serves me as my guide.
This Army shocks our enemies,
Bringing them crashing to their knees.

Basic Training is plenty rough—
To make it through you must be tough.
Squad Leader, don't be blue,
They're gonna make you a soldier, too.

Victory Academy, U.S. Army Drill Sergeant School, Fort Jackson, SC

COMPANY C

I'm gonna fill me a heavy, heavy pack,
Filled full of concrete, fill it for the trip.
When you enter my tent, be prepared to die,
Because I'm hand-to-hand qualified.
I don't care what they say,
Proud to be from the U.S.A.
Fright for freedom and fight for fun,
U.S. Army is Number One!

Company C, 16th Ord. Bn., Aberdeen Proving Gnd, MD

DOUBLE TIME RUN

Double time, double time up the hill,
Everybody's gonna get a two mile thrill.
Double time, double time, everyone will,
Everybody gonna get their fill.

Double time, double time, two miles long,
How in the hell can we go wrong?
Double time, double time, to this song,
Everybody's gonna make it strong.

Double time, double time, going strong,
Who ever thought we could run so long?
Double time, double time, all the way,
We get up and run all day.
Double time, double time, love it, too.
We can run the whole day through.
Double time, double time, everybody's havin' fun,
We can't wait for the four mile run!

Victory Academy, U.S. Army Drill Sergeant School, Fort Jackson, SC

OLD LADY DIVER

Saw an old lady walkin' down the street,
She had tanks on her back, fins on her feet.
I said, "Hey, old lady, where're you goin' to?"
She said, "U.S. Navy Diving School!"
I said, "Hey, old lady, ain't you been told?
You'd better leave the scuba divin' to the brave and bold."
She said, "Sonny, sonny, can't you see? I teach Recon, U.D.T!"

Anonymous, U.S. Marine Corps, Quantico, VA

WAC SOUND OFF

*Note: use regular **Sound Off!** chorus in between verses.*

Joined the Army to wear my greens,
All I do is clean latrines.

CHORUS
Use the Sound Off! chorus.

I got a guy in old Milwaukee—
We make love by walkie-talkie!
(Repeat chorus)

I got a guy in New York City,
So cross-eyed he thinks I'm pretty!
(Repeat chorus)

Big black shoes and long blue skirt,
Gee, now, you look like a jerk!
(Repeat chorus)

I don't know but I hear rumors,
SGT _____ wears khaki bloomers.
(Repeat chorus)

Trainee, trainee, don't be blue,
Our Recruiter fooled us, too!
(Repeat chorus)

Joined the Army to wear my brass,
But all I do is ... cut the grass.
(Repeat chorus)

M. Cornell, Fort Lewis, WA

WHAT THE ARMY'S DONE TO ME

Joined the Army to get a degree,
And now I've got my Ph.D.

CHORUS
Oh, oh, oh, oh!
Oh, oh, we've got to go!
Oh, oh, oh, yeah—
Oh, oh, oh, oh!

Momma, Momma, can't you see?
What this Army's doing' to me!
(Repeat chorus)

They took away my faded jeans,
Now I'm wearin' Army green.
(Repeat chorus)

They took away my gin and rum,
Now I'm up before the sun.
(Repeat chorus)

Thought I'd get to have some fun,
Now all I do is shoot my gun.
(Repeat chorus)

Joined the Army to get in shape,
Now all I do is hurry and wait.
(Repeat chorus)

Used to drive a Chevrolet,
Now I'm walking all the way.
(Repeat chorus)

Used to drive a Cadillac,
Now I pack it on my back.
(Repeat chorus)

Captain, Captain, can't you see?
What this Army's doin' to me.
(Repeat chorus)

Soldier, soldier, don't be blue,
This Army's gonna take care of you.
(Repeat chorus)

Victory Academy, U.S. Army Drill Sergeant School, Fort Jackson, SC

NIGHT FLIGHT

Jump in, strap in, head for the sky,
Cobras flying in the night.
Dive and climb throughout the flight,
Take them in and keep 'em tight.

Keep the target in the sights,
Grab the trigger and squeeze it tight.
Bullets flying through the night,
Job is done and we're all right.

SP5 Pferdner, 6th Bn., Fort Rucker, AL

TWO YOUNG MEN

Two young men from Birmingham,
Went to talk to Uncle Sam.

One drove a Caddy with a sun roof top,
Now he's doin' the P.T. rock.

The other one came in an Oldsmobile,
Went infantry, don't need no wheels.

They made it through Basic with their hands held high,
Hit Airborne School and almost died.

Now there's wings upon their chests,
From a lotta time in the lean and rest.

Three years went by, their time was done,
One stuck around, he was havin' fun.

The other one seemed to be in shock,
He took the closest route back to the block.

SSG Wm. A. Honeycutt, 331st Trans. Co., (ACV)2, Fort Story, VA

UP JUMPED A SOLDIER

Up jumped a soldier from a powder pit,
Said "When it comes to fightin', you know I'm fit!"

He lined a hundred fighters up against a wall,
Said "I'll bet five dollars I can beat 'em all!"

Well, he beat ninety-seven, then he fell to one knee,
Looked around, turned around, beat the other three.

When he died, he went straight to Hell,
He beat up the Devil and some demons as well.

On his tombstone, written in green,
It says "Here lies a Hard-chargin' Soldier Machine!"

CPT M. Combest, C CO., 1/3 FA 2nd Arm. Div., Fort Hood, TX

NIGHT FLIGHT

Jump in, strap in, head for the sky,
Cobras flying in the night.
Dive and climb throughout the flight,
Take them in and keep 'em tight.

Keep the target in the sights,
Grab the trigger and squeeze it tight.
Bullets flying through the night,
Job is done and we're all right.

SP5 Pferdner, 6th Bn., Fort Rucker, AL

BIBLE STORIES

Jesus was an Airborne Ranger—
You'll be one too, oh Lordy.
Jesus was an Airborne Ranger—
You'll be one too, oh Lordy.
Jesus was an Airborne Ranger—
You'll be one too, oh Lordy.

CHORUS
Look away beyond the blue horizon!

Adam was a Basic Trainee—
You will be too, oh Lordy.
(repeat above verse twice)
Look away beyond the blue horizon!

Adam and Eve wore O.D. green—
You'll wear it too, oh Lordy.
(repeat above verse twice)
Look away beyond the blue horizon!

Cain was an "Abel" killer—
You'll be one too, oh Lordy.
(repeat above verse twice)
Look away beyond the blue horizon!

Delilah was an Army barber—
You'll meet one too, oh Lordy.
(repeat above verse twice)
Look away beyond the blue horizon!

Sampson got a G.I. haircut—
You'll get one too, oh Lordy.
(repeat above verse twice)
Look away beyond the blue horizon!

CHAPTER 3: General Cadences

Noah was in Transportation—
You'll be there too, oh Lordy.
(repeat above verse twice)
Look away beyond the blue horizon!

Noah made an amphibious landing—
You'll make one too, oh Lordy.
(repeat above verse twice)
Look away beyond the blue horizon!

Moses low-crawled up the mountain—
You'll low-crawl too, oh Lordy.
(repeat above verse twice)
Look away beyond the blue horizon!

Moses was a lost lieutenant—
You'll be one too, oh Lordy.
(repeat above verse twice)
Look away beyond the blue horizon!

Moses got some regulations—
You'll get some too, oh Lordy.
(repeat above verse twice)
Look away beyond the blue horizon!

Pharaoh was a tank commander—
You'll be one too, oh Lordy.
(repeat above verse twice)
Look away beyond the blue horizon!

Pharaoh was a Navy diver—
You'll be one too, oh Lordy.
(repeat above verse twice)
Look away beyond the blue horizon!

Caesar was a conquering hero—
You'll be one too, oh Lordy.
(repeat above verse twice)
Look away beyond the blue horizon!

Ben Hur ran the motor pool—
You'll be one too, oh Lordy.
(repeat above verse twice)
Look away beyond the blue horizon!

Jesus ran a twelve man A-team—
You'll lead one too, oh Lordy.
(repeat above verse twice)
Look away beyond the blue horizon!

John the Baptist was court-martialed—
You will be too, oh Lordy.
(repeat above verse twice)
Look away beyond the blue horizon,

Peter was an ear lobe slicer—
You'll slice some too, oh Lordy.
(repeat above verse twice)
Look away beyond the blue horizon!

Lazarus worked the graveyard shift—
You'll work it too, oh Lordy.
(repeat above verse twice)
Look away beyond the blue horizon!

Thomas doubted regulations—
You'll doubt them too, oh Lordy.
(repeat above verse twice)
Look away beyond the blue horizon!

Jehovah was a ten-star general—
You'll be one too, oh Lordy.
(repeat above verse twice)
Look away beyond the blue horizon!

Satan was a fire-team leader—
You'll be one too, oh Lordy.
(repeat above verse twice)
Look away beyond the blue horizon!

Satan went and got demoted—
You'll do it too, oh Lordy.
(repeat above verse twice)
Look away beyond the blue horizon!

CDT Pattie Reichle, YSU, OH

RECRUIT

Recruit, Recruit, can't you see?
Being mean is killing me.
Drill Instructor, all I see,
Is that you sure don't look sad to me.

Recruit, Recruit, I sure feel bad,
Walkin', actin' like I'm mad.
Drill Instructor, all I see,
Is that it doesn't look like an act to me.

Recruit, Recruit, where have you been?
Round this island and back again.
Whatcha gonna do when you get back?
Gonna take a shower and hit the rack!

U.S. Marine Corps Recruit Depot, Parris Island, SC

AIR FORCE BASIC

Basic, Basic, don't feel blue,
Six more weeks and you'll be through.
When you get there, you will know,
The Air Force is the way to go!

Look up, look up in the sky,
F-15 goes flying by.
Joined the Force to wear the Blue,
So I can fly an Eagle, too!
(Repeat chorus)

Anonymous, U. S. Air Force Academy, Colorado Springs, CO

TINY BUBBLES

Tiny bubbles, in my beer,
Make me happy, while I'm here.

Tiny bubbles, in my glass,
Make me happy, full of sass.

Tiny bubbles, to the top,
I'm so full, I could pop!

Anonymous, Fort Bragg, NC

SCHOFIELD

Jody don't live in this here place,
He can't stand the heavy pace.

CHORUS
Back on the rock,
Marchin' up the block!

Even though the sky is blue,
We work too hard—this is true!
(Repeat chorus)

Hula girls and big palm trees,
You can't see them on K.P.
(Repeat chorus)

Prices are high and cash is low,
Savings don't get a chance to grow!
(Repeat chorus)

But all-in-all, I like it here,
Magnum does—that's real clear!
(Repeat chorus)

Anonymous, Schofield Barracks, HI

NAVY

Air Force says that they fly high,
We say they don't even try.
If you want the very best,
Navy puts you through the test.

Army ain't got nothin' new,
Marines say they're proud and true.
But if you want the very best,
Navy puts you through the test.

Hey, troops, get out of bed!
Get your a__ (butt?) 'n' into that head.
Get those pea coats off the rack,
Before the dawn can even crack.

Used to stay up drinkin' wine,
Now I have my shoes to shine.
Bunks and drills ain't no big thing,
As long as I get to sing.

C.C.s try to make us shine,
C.C.s say that we are fine.
Give us something else to do,
We'll do that plus something too!

Anonymous, Naval Training Center,
Great Lakes, IL

AIR FORCE

F-15 rolling down the strip,
Eagle driver gonna take a little trip.

Rev it, taxi up, count to four,
Push the throttle forward and hear the engines roar.

Thirty thousand feet up in the air,
Flying this baby is a natural high.

Took a look at six o'clock and what did I see?
A MIG-21 was comin' after me.

Pulled it up, rolled it out, much to his surprise,
Should've seen the look in that turkey's eyes.

Got behind him, set my sights, let my missile fly,
Blew that twenty-one outta the sky.

When you see an Eagle driver, he will say,
"Flyin' and fightin' is the Air Force way!"

Anonymous, U.S. Air Force Academy, Colorado Springs, CO

IV

We compete to prove we're fit
We always march and never sit.
You should see the way we run,
We can prove we're Number One!

Competition and Rivalry

HEY, HEY!

Hey, Hey, Hey, Hey!
Running this way every day.

Ho, Ho, Ho, Ho!
Bravo knows the way to go.

Hee, Hee, Hee, Hee!
Delta never will catch me.

Hoo, Hoo, Hoo, Hoo!
Neither will Alpha, neither will you!

Anonymous, PAO, Fort Gordon, GA

AIR FORCE PRIDE

I'm not sure, but I think I see,
A bunch of Army goons just lookin' at me.

I'm marchin' so fast and struttin' my stuff,
They were jealous as hell and left in a huff.

Anonymous, U.S. Air Force, Lackland Air Base, TX

IF I WAS PRESIDENT

If I was President,
And had my way.
There wouldn't be a "leg,"
In the Army today.

I'd box 'em up, wrap 'em up,
Then send 'em on home.
'Cause up in the sky,
Is a-where I roam.

SPC Patrick Kelley, D. Btry. 1/41 FA 56th Bde (Pershing), Germany

SINBAD

Sinbad sailed the Seven Seas,
And so do we to keep you free.
Our ships are mighty, our guns shoot straight,
Navy pushes by when the grunts are late.
ic-tor-reee is our solemn vow,
Navy brats will teach you how!

Great Lakes Naval Training Center, IL

THE FIGHTING GREEN MACHINE

We're the fighting Green Machine,
Better than any old Marine.
Fightin' is our middle name,
Honor, valor is our game.

Fightin' men one and all,
Never one to stop or fall.
Always there to make it right,
Never running from a fight.

If trouble comes we get the call,
We're fightin' men, one and all.
In far-off lands or native shore,
We always try to give much more.

We never falter, we never hide,
We always show our Army pride.
Airmen may be proud of blue,
But we can fly their damn planes, too!

Marines, be proud of your few,
Cause we can do things you can't do.
We're always at our country's call,
We're fighting soldiers, one and all.

MSG D. N. Rodriguez, HHC, DISCOM, Fort Polk, LA

FRANKFURT'S NUMBER ONE

Frankfurt's soldiers are Number One,
We run two miles just for fun.
Make three miles, it's getting light,
We run until we're out of sight.

Hey, hey, Heilbronn, you can hide your face,
Because Frankfurt will set the pace.
Mainz can hide 'em, too,
Because Frankfurt's gonna breeze on by you!

Office of CSM, 3rd Infantry Div.

FOLLOW ME

Hey-ey there, Army,
Get in your tanks and follow me.
I am the Marine Corps Infantry!

Hey-ey there, Navy,
Get in your ships and follow me.
I am the Marine Corps Infantry!

Hey-ey there, Air Force,
Get in your planes and follow me.
I am the Marine Corps Infantry!

Hey-ey there, Civilians,
Get off your butts and follow me.
I am the Marine Corps Infantry!

U.S. Marine Corps Recruit Depot, Parris Island, SC

AIR CAV TROOPER

Hey there, Jody, have you heard?
I'm gonna crew a mean war bird.
Air Cav Trooper in the sky,
Man, you know I'm gonna fly.

Cobra-TOW is mighty fine,
Through those trees those rotors whine.
Snake's gotta have its ears and eyes,
That is why the Scout crew flies.

O-H Fifty-eight can glide,
hrough the trees, down the mountain side.
Sneak and peek, never hesitate—
Tanks never see him 'til it's too late.

Scout finds the tanks, calls the Cobra down,
Snake slides along just above the ground.
Up through the trees, it's a deadly thing—
Tankers gonna die when they feel its sting.

I like it here on the Air Cav side,
My trade mark is "All Guts and Pride."
Can you do it? Can you pass the test?
And be like me, "above the best!"

CPT J.B. Norwood, 41st Co, 4th Bn., Fort Rucker, AL

IF YOU WANT TO BE

If you want to be a doggie,
Then you should've joined the Army!
Hey, hey, yah!
Hey, hey, yo!

If you want to be a swabbie,
Then you should've joined the Navy!
Hey, hey, yah!
Hey, hey, yo!

If you want to be a fly boy;
Then you should've joined the Air Force.
Hey, hey, yah!
Hey, hey, yo!

But if you want to be in the Marine Corps,
So be glad you joined the Marine Corps.
Hey, hey, yah!
Hey, hey, yo!

If you want to be an Officer,
Then you gotta go to Quantico.
Hey, hey, yah!
Hey, hey, yo!

But if you want to be a D.I.,
Then you have to go to San Diego.
Hey, hey, yah!
Hey, hey, yo!

Drill Instructor Chant Book, U.S. Marine Corps Recruitment Depot, San Diego, CA

I'M NOT SURE

I'm not sure, but I think I smell,
A boat full of Navy men, but it's too far to tell.
We're running so far, and it's not even five,
If they'd open their eyes, they'd see us jive.

I'm not sure, but I think I hear,
And Air Force jockey flying in real near.
From where he sits we look like ants,
Marching in a pack and singing these chants.

Anonymous, Fort Lewis, WA

AIR FORCE BLUE

Air Force, Air Force, we come through!
We fly on by, laughin' at you.

The clothes we wear are blue like the sky,
The thing we do best 'course is fly.

Anonymous, U.S. Air Force, Ramstein, Germany

PRAISE

Alpha Company, lining up along the avenue,
All they want to do is praise me and you.

So, praise the Lord and keep the soldiers coming,
C-4-3 is stepping out tonight, oh, baby!

Drill Sergeants of C-3-3, Fort Dix, NJ

ARMY, NAVY

Army, Navy, where you been?
Down to P.I. and back again.
Army, Navy, what did you see?
D.I. students runnin' P.T.
Singing left, right, left ...

Army, Navy, what did you do?
Well, I jumped right in and P.T.'d, too.
Singing left, right, left ...

Army, Navy, what did you think?
Well, I think I want to be a U.S. Marine.
They sing left, right, left ...

Air Force, Air Force, don't be blue,
The Marines are gonna teach you how to
P.T., too.
We sing left, right, left!

U.S. Marine Corps Recruit Depot, Parris Island, SC

FLYING

Flying high and flying low,
Watching the infantry on the go.
Flying in and flying out,
Here the Infantry scream and shout.

Flying up and flying down,
I'll gig the enemy, shootin' up the ground.
Flying forward and flying back,
I'll help the grunts in their attack.

U.S. Marine Corps Recruit Depot, Parris Island, SC

CO. 401

Hey, yo, Captain Brown,
4-0-1 don't monkey around.
We're here to be the very best,
To put our mark on the great Midwest.

CHORUS
Hidi, hidi, hidi, hey!
Hody, ody, ody, ho!

Some of us may be old,
But now we're solid gold.
Some of us, we are young,
But 4-0-1 is having fun.
(Repeat chorus)

If you see us on the street,
You will know we can't be beat.
We will be the best around,
4-0-1 don't monkey around.
(Repeat chorus)

Co. 401, Naval Training Center, Great Lakes, IL

U. S. MARINE PRIDE

The Army and the Navy went to Vietnam,
But they couldn't quite handle the Viet Cong.

Marines went in to pull them out,
We filled Old Charlie full of doubt.

There were a lot of wounded, and a lot of dead,
But most of them were the Zipperheads!

U.S. Marine Corps Recruit Depot, Parris Island, SC

GOLDEN HAWK

Hawks we know, are birds of grace,
They have the right to all airspace.
You see them fly with speed and heat,
If you move too slow they'll snag your feet!

These are words of mighty wit,
When you see 'em comin', you'd better git!
Bringing death from high above,
We hawks are swift, we show no love!

Now if you think that it's a lie,
Try your luck and it's bye-bye!
We're on the go and above the rest,
The Golden Hawks will never rest!

*SPC Sample, 14th Co. 1st Av. Bde, 2nd Ed. Jody Calls,
Fort Rucker, AL*

EXTRA

Extra! Extra! Hot from the press—
Navy saves the Army from a great big mess.

Extra! Extra! Hear the news—
Navy comes through, Army is blue.

Extra! Extra! It ain't no jive—
Navy keeps those Army boys alive.

U. S. Navy Training Center, Great Lakes, IL

CO. 392

Get off the grinder, get on the grass,
3-9-2 is marching past.
Get off the grinder, get on the green,
3-9-2 is looking mean!
Am I right or wrong—you're right!
Are we weak or strong—we're strong!
Sound off!
U.S.
Sound off!
U.S.
Navy!
Bring it on down
U.S. Navy, U.S. Navy, rah!

Co. 392, Naval Training Center, Great Lakes, IL

MARINE TOUGH

Groovy, groovy, groovy,
Tough, tough, tough.
You might beat everybody,
But you can't beat us!

Running, running, running,
Running every day.
We'll come out Number One,
Because we lead the way!

Drilling, drilling, drilling,
Drilling every day.
We'll win the final drill,
Because we lead the way!

Anonymous, U.S. Marine Corps, Quantico, VA

MARINE

You can keep your Army khaki,
And your Navy blues.
Because there's a different kind of fightin' man,
I'll introduce you to.

His uniform is different,
The finest ever seen.
The Germans called him Devil Dogs,
His real name is Marine!

He was born at Parris Island,
The land that God forgot.
The sand was eighteen inches deep,
The sun was blazing hot.

He'd walk a hundred miles or more—
Before the day is done.
The women call him 'macho man,'
His real name is Marine!

U.S. Marine Corps, Quantico, VA

TEAM SPIRIT

We fly to Korea 'most every year,
When it comes to Team Spirit, we have no peer.

We shop in Seoul while we're there,
Breathin' in all that kimchi air.

We walk down the streets all lined with shops,
Those drinkin' bars, they sure are tops.

We're Mad Dogs, don't get in our way!
Or your little ole yobos won't get paid.

25th Infantry Div., Schofield Barracks, HI

BIRDIE

Birdie, Birdie, in the sky,
I'm gonna teach you how to fly.
With cyclic pedals, collective, too,
And rotors flapping, just like you.
Hovering,
In ground effect,
Hovering,
In ground effect.
And taking off,
And taking off.

Nap-of-the-earth and high recon,
Birdie, this will turn you on.
With V.O.R. and I.L.S.,
Now, Birdie, fly back to your nest.

Office of the CSM, 3rd Infantry Div.

Co. 398

Hey, Air Force!
Pick up your jets and follow me.
We are the Navy, can't you see?

Hey, Marine Corps!
Pick up your rifle and follow me.
We are the Navy, can't you see?

Hey, Great Lakes,
We are Company 3-9-8.
We will be going through your gates,
With or without your seaman rates!

Co. 398, U.S. Naval Training Center, Great Lakes, IL

ALL THE WAY

We're the ones who run the most,
And that is not an idle boast.

Running makes our muscles strong,
Also makes the day seem long.

Ex-Er-Size is where it's at,
Body building so we ain't fat.

Look at _____ troop sittin' by the road,
They're so weak they run like toads.

If we had to we could run,
All the way to K-town just for fun.

Anonymous, Mannheim, Germany

COMPANY 3-6-7

Oh, you'd better not shout, you'd better not pout,
You'd better not cry, I'm telling you why!
3 - 6 - 7 is marching by.

You'll see us when you're sleeping,
You'll know that we are great!
You'll know that we're the greatest,
Company on this Naval base!

Company 367, U.S. Naval Training Center, Great Lakes, IL

WAY OUT FRONT

Way out front, tell me what do you see?
Scout Platoon from C-S-C.
A-T, A-T, G-S-R Won't be behind them very far.
_____ Platoon will be in the rear,
Bringing in smoke from far and near.
_____ Platoon will do the best—
They support us with the rest.
Way out front, tell me what do you see?
Bear track, bear track looking back at me.
Must be Bravo, the King of the Bear,
Leaving that bear stuff everywhere.
They're big around the middle, small across
  the rump.
They can run but sure can't jump.

SGT R. Smith, Fort Lewis, WA

EVERYWHERE WE GO (U.S. Army version)

Everywhere we go, people want to know,
Who we are—so we tell 'em:

We are Bravo, Big Bad Bravo!
Better than Alpha, awful-awful Alpha!

Better than Charlie, chicken-chicken Charlie!
Better than Delta, dumb-dumb Delta!

Better than Echo, icky-icky Echo!
'Cause we're Bravo!
Rough, tough Bravo,
Lean, mean Bravo, Yeh!

SGT Max Woodlee (Ret.), Fort Lewis, WA

AVIATOR

Aviator, aviator, what do you see?
An Infantry unit under me.

Aviator, aviator, what're you going to do?
Pick some up and get the rest through.

Aviator, fire your missile on a wire.
Gonna give our boys some suppressive fire.

Aviator, aviator, above the rest.
We'll fly north, south, east, 'n' west.
If an enemy round should stop your climb,
Hit your collective and land on a dime.

Orange Flight 79-5, 1st Ed. Jody Calls, Fort Rucker, AL

CHAPTER 4: Competition & Rivalry 111

EVERYWHERE WE GO (U.S. Navy version)

Everywhere we go, people want to know,
Who we are—so we tell 'em:

We're not the Army, the back-packing Army!
We're not the Air Force, the low-flying Air Force!

We're not the Marines, they don't even look mean!
We're not the Coast Guard, they never, ever work hard!

We are the Navy, the world's finest Navy!
(Leader)
    Am I right or wrong?
(Troops)
    You're right!
(Leader)
    Are we weak or strong?
(Troops)
    We're strong!
(Leader)
    Sound off—
(Troops)
    We are!
(Leader)
    Sound off—
(Troops)
    Awesome!
(Leader)
    Bring it on down—
(Troops)
    We are awesome! We are awesome, OOSH!

Anonymous, U.S. Naval Training Center, Great Lakes, IL

ALPHABET

A-B, C-D-E, Delta Company's where I want to be,
F-G, H-I-J, look out Army 'cause I'm here to stay.
My boots are clean, they're shining bright,
I know I'm looking out of sight.
Who's that, who's that?
Who's that looking so bad? Hey!
Delta, Delta, Delta, Delta is sounding so bad.

K-L, M-N-O, everybody is wanting to know,
Who's the company out on the go?
P-Q, R-S-T, everybody wants to be like me,
This is Delta Company, can't you see?
Who's that, who's that?
Who's that sounding so bad. Hey!
Delta, Delta, Delta, Delta is sounding so bad!

SFC J. Gray, ROTC Instructor, Radford, VA

MARINES VS. ARMY

I don't know, but I've been told,
The Marine Corps thinks it's mighty bold.
They don't know what the Army can do,
We are proud of our history, too.

Our looks and style may not be smooth,
But man, you oughta see this Army move!
Look to your left, and what do you see?
A bunch of jarheads looking at me.

Sing it out and sing it loud,
I'm a soldier and I'm mighty proud!

Anonymous, Fort Benning, GA

V

Am I right or am I wrong?
Each of us is tough and strong.
We guard the ground, the sea, the sky,
Ready to fight, willing to die.

Special Selections

This chapter contains lots of sharp selections for Armor, Artillery, Infantry, Airborne, Aviation, and Air Assault units.

THE SPIRIT OF THE CAVALRY

Scouts out front and mortars to the rear,
Shillelagh missiles gonna make 'em fear.
No other Unit you'll ever see,
Has the spirit of the Cavalry!
Load up, mount up, move to the field,
Cavalry is the Army's shield.
No other Unit could ever have,
The fighting spirit of the Cav!

Daytime, nighttime, anytime at all,
The Cav can answer freedom's call.
It ain't no bull, it ain't no lie,
The Spirit of the Cav will never die!

Troop E, 2nd Squad, 6th Cavalry, **Jody Calls**, *Fort Knox, KY*

GUNNER'S CHANT

I was born with a lanyard in my hand,
I'm a real straight shooter, I'm a gunnin' man.
They call me Cannon Cocker, and I'm Number One,
I'm a fifty-five baby, I'm a son-of-a-gun!

Give me a 'jo and a long tube gun,
And I'll put the enemy on the run.
Ain't nobody as bad as me,
I'm First! of the Third Artillery.

CPT M. Combest, C. Com. 1/3rd FA, 2nd Armored Div., 1st Ed. **Jody Calls,** *Fort Hood, TX*

HISTORY LESSON

Gunners, gunners, listen to me,
While I tell you about the 3rd Artillery.
Since we began in 1794,
No other artillery battalion has ever done more.
In 1812 we bloodied the British at Washington,
Fifty years later we fought at Bull Run.
Steadfast and true during the war with Mexico,
We lost, then regained our colors against the Seminole.

We were there if France during World War One,
Before that, we fired on Peking during the Boxer Rebellion.

Gunners, gunners, listen to me,
True to our cannons we'll always be.
Gunners, gunners, can't you see?
No one is better than you and me.

CPT M. Combest, C. Co. 1/3 FA, 2nd Armored Div., **Jody Calls***, Fort Hood, TX*

REDLEG RUNNING SONG

Do you job now, 'n' do it right,
You'll go home, see Momma tonight.
Mess around, don't do your work,
Call you in, call you a jerk!
Artillery is the branch for me,
Other guys go to the Infantry.

Grunts run hard and they walk far,
In Artillery we ride by car.
Gunner, A.G., Number One!
All cannoneers, we like th' big guns.

Close the breech and pull the string,
Man, that oh-five sure can sing.
F.O. says first rounds look bad,
Check your sights, see what you had.

F.D.C. will slip their sticks,
Two more rounds now, make it quick!
F.O. says those rounds were close,
Six more rounds, we'll drink a toast.

F.D.C. sends a new Q.E.
Fire for effect with the Battery.
F.O. says those rounds were fine,
Let's go home and drink some wine.

One-oh-five goin' down the range,
Red hot steel makes you feel strange.
Hit you high and hit you low,
Try to hide, ain't nowhere to go.
Here we go, all the way,
All the way, Redleg!

SSG Caldweld, 260th FAD, 1st Av. Bde, 2nd Ed. Jody Calls, Fort Rucker, AL

ARTILLERY

Mama, oh, Mama, now can you see?
They put me in the Artillery.
Men of war with their heads held high,
They're ready to fight, they're willing to die.

Papa, oh, Papa, say have you heard?
The General went and put me down in Third.
They've got a fist of iron and nerves of steel,
If the "quick" don't get you, then the "Artys" will.

Tell my baby to forget about me,
I've got a new lover, the Artillery.
Tell my baby, don't feel blue,
Tell her t'join the Artillery, too!

CPT M. Combest, C. Com. 1/3 FA 2nd Armored Div., 1st Ed. Jody Calls, Fort Hood, TX

FIELD ARTILLERY

M-1-0-9 rolling' down the trail,
Look at our enemies turnin' tail.
H-E 1-C-M rainin' down,
W-P burnin' all around.

If these shells are not enough,
Then the artillery will have to get rough.
If the enemy wants to die,
Atomic wisdom is our reply.

Watch 1-0-9 as it flies a round,
Feel the impact shake the ground.
Suppression, destruction, obscuration, too,
There isn't much that we can't do.

It's our preps and deadly fire,
That is maneuvers great desire.
Mixed with Infantry, planes and tanks,
We're the punch in the combined arms ranks.

1/78 Field Artillery, 2nd Armored Div., 1st Ed. Jody Calls, Fort Hood, TX

OCS

Hail, hail O-C-S,
You look at us - you see the best.
Hail, hail Infantry,
Queen of Battle, follow me.
Up in the morning and out of the rack,
You're greeted by your friendly TAC.
Drop down—give me ten!
Too slow—do it again!
Hail, Hail O-C-S,
You look at us - you see the best.
Hail, hail Infantry,
Grab your rifle, follow me.

J. H. Lutz, Toronto, Canada. TAC, Fort Benning, GA

ARMOR

Up in the morning at the crack of dawn,
Shined my shoes 'fore I put 'em on.
Day uniform is squared away,
Now I'm ready to start my day.

CHORUS
One for the money, two for the show,
Three sixty-six is ready, and we're on the go!

Took my M-16 to the G-3 test,
Did real good, but not my best.
I cleaned my chamber and the bore,
Now I'm ready for the game called War.
(Repeat chorus)

Sergeant said, "You're in lane four,
Set that lever on 'rock 'n' roll',
Zero on the target and knock it down,
Watch out for that hot brass round!"
(Repeat chorus)

All the way it's a ten mile march,
First Sergeant said we would walk, walk, walk.
We walked all day and part of the night,
When I hit the sack, I'll be feelin' all right!
(Repeat chorus)

PFC G. F. Napier, HHC, 3/66 Armor, 2D Armored Div., 1st Ed. Jody Calls, Fort Hood, TX

2/67 ARMOR

Bo Didley, Bo Didley, have you heard?
Can't drop a tank from a big iron bird.
Airborne Ranger's nice I guess,
But we all know that tankin's best.

When we roll by, Airborne salutes,
'Cause they can't stop hordes with their parachutes.
The work is done by guys like me,
And poor souls in the Infantry.

Bo Didley, Bo Didley, when they come,
I've got a 1-0-5 tank gun!
Button them up with artillery,
They'll be so blind they can't see me.

Engineers will come and help us, too,
Ditches and mines, they'll give a few.
I wish I may, I wish I might,
Get the rest with battle sight.

Bo Didley, Bo Didley, life is great,
Serving here in the Lone Star State.
If duty calls then I will go,
Wherever I must to fight the foe.

To the Middle East or Europe's shores,
To serve my country in the wars.
Just so everybody will understand,
I'm a 2/67 fightin' man.

CPT Douglas Boulter, 2d Armored Div., Fort Hood, TX

THESE BOOTS—ARMOR

These boots were made for walkin'
And that's just what they'll do.
If you pick a fight with us,
They'll walk all over you.

These tanks were made for drivin'
And that's just what they'll do.
Give us half a reason,
And we'll drive all over you.

The Cav was made for ridin'
And that's just what we'll do.
If you need a bruisin'
We'll cruise all over you.

These tracks were made for layin'
And that's just what they'll do.
You'd better start a-prayin'
Or we'll put 'em down on you.

These guns were made for shootin'
And that's just what they'll do.
If we get a mission,
We'll drill a hole in you.

These birds were made for flyin'
And that's just what they'll do.
If you keep on lyin'
We'll fly 'em over you.

This Army's made for fightin'
And that's just what we'll do.
If you pick a fight with us,
We'll walk all over you!

J. McClelland, ***Jody Calls****, Fort Knox, KY*

ARMOR IS A WALL OF STEEL

Armor is a wall of steel,
That faces every foe.
No kin can find an equal,
To Armor's mighty role.

Tradition is another name,
For Cav's illustrious creed.
The Cavalry's ever noted,
In Army's foremost deeds.

Just as we proudly choose,
Our nation's honored banner.
So Armor is our place to serve,
In the Army's highest manner.

Our country is our duty,
Our people are our trust.
We'll always go with Armor,
To end dictators' lusts.

J.L. Molloy, ***Jody Calls****, Fort Knox, KY*

MECHANIZED WARRIORS

One-one-three rollin' down the trail,
Mechanized Warriors made of solid steel.

Headspace set and timing's just right,
The fiftys gunner's ready and we're looking
for a fight.

Contact made, we maneuver to the right,
Wouldn't you know it's the middle of the night.

Action's heavy, but we've seen worse,
Straight, Stalwart, Mechanized, One-Forty-First.

The sun finally shows what we all know,
They sent their best, and failed like the rest.

PFC Sankey, B.Co., 1/41 Inf, 2d Armored Div., 1st Ed. Jody Calls, Fort Hood, TX

MECHANIZED INFANTRY #1

I don't know, but I've been told,
Dolphin's Spirit's good as gold.
Run all day and fight all night,
Up again at first daylight.

We don't mind it, no way,
This is the way to start our day.
Dodging bullets and chewing nails,
Bravo Company's tough as Hell.

1-1-3s roll across the land,
Mechanized grunts gonna make a stand.
Overwatch your buddy to the bottom of the hill,
Attack the objective and make that kill.

Mobile foxholes are the only way,
Cover sixty miles in a single day.
Easy on your feet and you'll learn to say,
Mechanized Infantry all the way!

Lt. Collier, B. Co., 1/41 Inf., 2nd Arm. Div., Ft Hood, TX

MECHANIZED INFANTRY #2

Gather 'round soldiers and I'll tell you a story,
About mechanized infantry in all its glory.
Let's start off the tale with the A-P-C,
Commonly known as the 1-1-3.

Now the 1-1-3 sure was fun,
Until the came out with the A-1.
Diesel Powered for get up and go,
But the A.P.C. was still too slow.

When suddenly there came in a gust of breath,
Twenty-five tons of sudden death.
Firing ports on the sides and the rear,
Help to control the soldier's fear.

Shoot an enemy from the track,
Send them all to hell and back!
The 25-millimeter and the TOW,
Can kill and destroy any type of foe.

The M2 Bradley is here to stay,
Mechanized Infantry—*All the Way!*

SSG G. L. Hakala, US Army Field Recruiter, Danvers, MA

INFANTRY

I say the field is pretty rough,
That's why the Army is mighty tough.
We're combat ready every day,
That's how we earn our monthly pay.

CHORUS
OH-OH-OH-OH
Oh Lord, I want to go!
OH-OH-OH-OH
Oh Lord, I want to go!

I say we train pretty hard,
Your safety we will guard.
People, people can't you see,
That we are the unit called Infantry.
(Repeat chorus)

Everyday we run PT,
They run the sweat right out of me.
I say we run a country's mile,
But we all finish with a great big smile.
(Repeat chorus)

You know this country's in a jam,
That's why we work for Uncle Sam.
To keep this country safe and free,
So we can live in Liberty.
(Repeat chorus)

SPC E. T. Glenn, HHC, 1st Bde, 2d Arm. Div., Fort Hood, TX

OLD SOLDIER

Old soldier, O-o-old soldier,
Pick up your rifle and follow me—
Fight with the Infantry.

Eighty-second, All-Americans,
Pack up your 'chute and follow me—
We are the Airborne Infantry.

One-oh-One, Screamin' Eagles,
Grab your weapon and follow me—
We support the Infantry.

Second Armor, Hell on Wheels,
Start your tanks and follow me—
We support the Infantry.

Fort Richardson, Northland Guardians,
Pick up your snowshoes and follow me—
We are the Arctic Infantry.

Ninth Division, Old Reliables,
Pick up a weapon and follow me—
Foot Cav soldiers, Infantry.

Old soldier, O-o-old soldier,
When I marry, my wife will be—
Airborne, Ranger, Infantry.

Old soldier, O-o-old soldier,
We will raise a family—
A whole platoon of Infantry.

CPT Petrone, SPO, Karlsruhe, Germany

197th INFANTRY BRIGADE

One-Ninety-Seventh is on the way,
Whether it be night or day.

All the Army says, oh, my!
When the One-Ninety-Seventh goes movin' by.

Heads held high, and pride within,
Whatever the mission, we're trained to win!

Nothing anyone will say or do,
Will ever make this unit number two.

We are mech; we're separate, too.
We're Forever Forward, through and through.

So if it's mech you want to be,
Say 1-9-7 Infantry!

CPT D. Bombaugh, 197 Inf. Bde. (Mech) (Sep), Fort Benning, GA

GOOD AS GOLD

I don't know, but I've been told,
Infantry blue is good as gold.
Work all day, play all night,
Infantry blue is fit to fight.

Dodging bullets, and chewing nails,
Infantry, Infantry, tough as hell.
Rucksacks, butt packs walkin' across the land,
Infantry grunts gonna make a stand.
Infantry blue is the best in the land!

SFC L. Joseph, 5th Unit, 3rd Bn., USA Correctional Activity, Fort Riley, KS

RANGER

Let the four winds blow,
Let it blow, let it blow.

From the east to the west,
Airborne Ranger is the best!

Hang your head, hang it low,
If Airborne you didn't go!

Sing it high, sing it low,
Sing it everywhere you go.

From the east to the west,
Airborne Ranger is the best!

D. Owings, A Co. Reserve Training Unit, Fort Lewis, WA

'CAUSE HE'S AIRBORNE RANGER, *ALL THE WAY*

Jesse James said before he died,
There's four things he wanted to ride.
Bicycle, tricycle, automobile,
A bow-legged horse or a ferris wheel.
'Cause he's Airborne Ranger, All the Way!

Down in the jungle where coconuts grow,
There's mean motor-scooter called Ranger Joe.
He's in all the way up to his knees—
He's an Airborne Ranger coming through the trees.
'Cause he's Airborne Ranger, All the Way!

Anonymous, Drill Sergeant School, Fort Dix, NJ

The following three Airborne chants of are typical examples of the changes that often occur to cadences over the years. Originally, the C-119 plane model was used; today a C-130 replaces it. It is also of interest that the popular book about soldiers in the *Vietnam War If I Die In A Combat Zone, Box Me Up And Send Me Home* (Tim O'Brien, Delacorte, 1973) is named after verses of Airborne cadences.

AIRBORNE #1

Stand up! Hook up! Stand in the door!
Stood up, then collapsed on the floor.

CHORUS
Use Sound Off! chorus.

Jumpmaster picked me up and then,
Stood me in the door again.
(Repeat chorus)
"Trooper do you mind the drop?"
"No, it's just that sudden stop!"
(Repeat chorus)
Jumpmaster tapped me out at last,
Jumped out in the old prop blast.
(Repeat chorus)
Fell on down, my mouth open wide,
Couldn't have counted if I tried.
(Repeat chorus)
Hit the ground with my feet apart,
In my stomach, felt my heart.
(Repeat chorus)
Started to drag and then I thought,
Cost me ten if I get caught.
(Repeat chorus)

Anonymous, Fort Bragg, NC

AIRBORNE #2

Soldier, soldier, have you heard?
I'm gonna jump from a big iron bird.

Up in the morning in a drizzling rain,
Packed my 'chute and boarded the plane.

Raining so hard that I couldn't see—
Jumpmaster said, "You can depend on me."

I looked with fear at the open door,
Then I stood up and fainted on the floor.

When I woke up, I was hooked up again,
And that is when I fainted again.

Anonymous, Fort Benning, GA

AIRBORNE #3

C-130 rollin' down the strip,
Airborne daddy gonna take a little trip.
When that plane gets up so high,
Airborne trooper gonna dance in the sky.

Stand up, hook up, shuffle to the door,
Started to jump but I fell on the floor.
They stood me up and pushed me to the door,
I jumped right out and counted to four.

If my main don't open wide,
I've got another one by my side.
If that one should fail me too,
Look out below, I'm comin' through.

Slip to your right and slip to your left,
Slip on down and do a P-L-F.
I hit the drop zone with my feet apart,
Feet in my stomach and legs in my heart.

Lying here and lying there, rollin' in fright,
Wonderin' if this is gonna be my last fight.
Nurse, oh, nurse, you look so fine,
Airborne trooper gotta make some time!

If I die in the old drop zone,
Box me up and ship me home.
Bury me in the leaning rest,
And tell all the girls I did my best.

If I die on a Korean hill,
Take my watch or the Commies will.
Put my wings upon my chest,
Tell the world I've done my best.

SGM W. Lawrence, Fort Lewis, WA

WHEN I GET TO HEAVEN

When I get to heaven,
Saint Peter's gonna say,
"How'd you make your livin,'
How'd you earn your way?"

I'll reply with a little bit of anger,
"I earned my way as an Airborne Ranger!
Blood, guts, tough and danger—
Death's no stranger to an Airborne Ranger!"

Anonymous, Fort Bragg, NC

SPELL AIRBORNE

A!   - All the way!
I!   - In the sky!
R!   - Running hard!
B!   - Born free!
O!   - On the run!
R!   - Runnin' hard!
N!   - Never stop!
E!   - Ever-ready Airborne!

On the run
In the sun,
One mile, no sweat,
Two miles, better yet.
Five miles, feelin' good,
Ought to be in Hollywood!

SFC J. Gray, ROTC Instructor, Radford, VA

I KNOW A GIRL

I know a girl who lives on the hill—
She won't do it, but her sister will.

What's this that her sister will do?
U.S. Army Airborne School!

D. Ownings, A Co. Reserve Training Unit, Fort Lewis, WA

AIRBORNE VS. STRAIGHT LEG

Airborne, "straight leg," which is best?
AIRBORNE! AIRBORNE! YES! YES! YES!
Airborne, "straight-leg," what-a-ya-say?
We're goin' Airborne all the way.
Battle cry of the "straight-leg" Corps—
Had some leave and need some more.

Anonymous, XVIII Airborne Corps, Fort Bragg,, NC

JUMP BOOTS

Coon skin and alligator hide,
Makes a pair of jumpboots just the right size.

Don't be afraid to put 'em on your feet,
My pair of jump boots can't be beat.

Shine 'em up, lace 'em up, put 'em on your feet,
Brand new jump boots just can't be beat!

U.S. Army Jump School, Airborne Department, Fort Benning, GA

PATHFINDER

Gonna tell you a story that is seldom told,
'Bout a mean Pathfinder, you can tell by his clothes.
Got a grease gun, K-Bar by his side,
These are weapons that he lives by.

U-H-1-H sittin' on the pad,
He gets on, sits down, lookin' bad.
Look up, standby, sit on the floor,
One-O, Two-O, Three-O, Four-O!

I slip to my right and I slip to my left,
I come down and do a P-L-F. Well,
I'm the baddest thing you've ever seen,
I'm an Airborne Pathfinder killing' machine.

Walking through the jungle with my M-16,
I'm the meanest thing that you've ever seen.
Up pops Charlie and what do you know,
I kick that soldier on a Rock and Roll!

1LT M. W. Sherman 14th Co., 1st Ed. Jody Calls, Fort Rucker, AL

STUDENT

Down the runway, headed high,
Comes a student about to fly.
When you look up in the sky,
A Rucker flier passes by.

Run and ring the warning bell,
It's a student, can't you tell?
They may graduate someday,
Then be entitled to that pay.

CPT Park, Class 78-44, Fort Rucker, AL

BOLD AVIATION

I don't know, but I've been told,
Aviators are mighty bold.
Look up high in the sky,
Huey gunships are flyin' by.

Aeroscouts are in the brush,
Cobra snakes are in a rush.
Pilots say it's killing me,
Waiting for the Infantry.

Flat-iron ships are very well read—
We go where others won't dare to tread.
Cranes go in to pick up the wreck,
Where the pilots have hit the deck.

Chinooks are coming with grunts in flight -
They want to hurry and join the fight.
Aviators are mighty bold,
That's the story I've been told.

R. A. Sample, 14th Co., 1st Bn., 1st Ed. Jody Calls, Fort Rucker, AL

UH-1

UH-1 hovering down the strip,
Airmobile Daddy gonna take a little trip.
Seat belt flappin' in the breeze,
Gonna N-O-E the Infantry.

UH-1 sitting on the pad,
Crew Chief's looking mighty sad.
For he knows he's gonna be sick,
Flying I-F-R in an out-dated slick!

1LT Tyson, 63rd Co., 1st Ed. Jody Calls, Fort Rucker, AL

AVIATION VARIETY

OH-58 is really great,
But all it is, is widow bait.

The "55" O-H, man, alive,
It's the meanest bird alive.

To fly a snake I sure do dread,
'Cause I know I'll end up dead.

To see my kids and wife is nice,
For me it's Hueys, pigs and rice.

The hook is big and ugly I know,
It's a trucker's only way to go.

Green Flight 78-73, 1st Ed. Jody Calls, Fort Rucker, AL

GREEN FLIGHT

Green Flight, Green Flight, how's your run?
Plenty tough, but lots of fun. Green Flight,
Green Flight, will you fly,
In those Hueys, by 'n' by?

Huey, Huey, where've you been?
Down to Cairns and back again.
Huey, Huey, why do you cry?
'Cause those WOCS, I gotta fly.

After nine months of those things,
Get our Warrants, get our wings.
With our wings upon our chests,
We will be "Above the Best!"

R. L. Scena, MSG. PAO, 1st Ed. Jody Calls, Fort Rucker, AL

AVIATION DADDY

U-H-1-H moving through the air,
Aviation daddy way up there.
Hooked up, buckled up, in the pilot's seat,
Having some fun at a thousand feet.

Hand on the cyclic, looking at the sky,
Co-pilot's worried about the way I fly.
Crew Chief's looking out the door,
Got seven Grunts starin' at the floor.

J. D. Moore, 62nd Co. (Royal Blue) Fort Rucker, AL
Moore was killed 11 December 1978.

TRANSITION

I started in a fifty-five,
Look out for that mutha in a dive!
To fly it is to me a joy,
It's more than just a Tinker Toy.
Am I right or wrong, this course is way too long,
Sound off, sound off, check your mags.
1, 2, 3, 4, 1, 2, —— 3-4!

Right now I'm at fifty-eight,
I wonder what I did to rate.
This offspring of a Huey, too,
Oh, well, to me it's something new.
Am I right or wrong, this will be my song,
Sound off, sound off, dump your nose.
1, 2, 3, 4, 1, 2, —— 3-4!

When I grow up I'd like to fly,
I long sleek Cobra through the sky.
I'd roll it out on Matteson Range,
Hit a steer or two just for a change.
Am I right or wrong, it will ring your gong,
Sound off, sound off, check your guns.
1, 2, 3, 4, 1, 2, —— 3-4!

And then when I get old and gray,
To drive Chinooks would be okay.
I'd do my thing without a fuss,
And show them it's not just a bus.
Am I right or wrong, straight and Level I belong,
Sound off, sound off, cargo check.
1, 2, 3, 4, 1, 2, —— 3-4!

Hanchey Div. Dept. of Flt. Training, 1st Ed. Jody Calls, Fort Rucker, AL

WOP WOP

Hey, everybody get on your feet,
Hey, everybody, listen to the beat.

CHORUS
Wop, Wop, Diddly Wop!
Wop, Wop, Wop!
Wop, Wop, Diddly Wop!
Wop, Wop, Wop!

Boots and saddles everyday,
This's the way we earn our pay!
(Repeat chorus)

Crank up the Cobras gettin' ready to fight,
Scouts in the day, Blue Teams at night!
(Repeat chorus)

W-O One hovering down the strip,
Everybody knows this's his first trip!
(Repeat chorus)

Hey, everybody, take a trip with me,
Down to Rucker where the rides are free!
(Repeat chorus)

MAJ T.A. Swindell, D. Troop, 2/1 Cav., 2nd Armored Div., Fort Hood, TX

64th COMPANY

Hey, Fort Rucker, have you heard?
I'm gonna fly me a whirly bird.
Preflight, buckle up, lock the door,
Rucker's I-Ps all ask for more.
Can't ever please 'em - always wrong,
So I do P-T and sing this song.

Hey, Colonel _____, have you heard?
I'm gonna fly me a whirly bird.
Six-Six-Hundred, rotor in the green,
Prettiest sight I've ever seen.
I'm gonna make it, filled with pride,
If I get through this next check ride.

Hey, Colonel _____, have you heard?
I'm gonna fly me a whirly bird.
Aviation training's Number One,
But this P.T. doesn't seem like fun.
If this don't kill me, can't you see?
Me and P.T.'s partin' company.

Hey, Captain _____, have you heard?
I'm gonna fly me a whirly bird.
All this working is crazy, it sounds,
To one who lifts up nine thousand pounds.
But we're gonna do it, don't you fret,
'Cause 64th Company's the very best yet!

ORWAC 78-36, 64th Co., 1st Ed. Jody Calls, Fort Rucker, AL

BLEW BLADES

Got together, talkin' with friends,
Talkin' about those stinkin' winds.

Sometimes headwinds, sometimes tails,
Mostly crosswinds, gusts like gales.

Pre-flight check, inspections due,
Engine deck, I found a worn out shoe.

Three dead birds in the air intake,
Maintenance said, "It's our coffee break."

Ninety knots, three thousand plus,
No hydraulics, don't make a fuss.

E.G.T. just went to nine,
Fire light's out - not a welcome sign.

Gear box out, we're spinnin' around,
Compressor stall, what a sound.

Sky's obscured, can't see my hand,
Wonder where I'll be forced to land.

Look at the Caution Panel, what do I see?
It's all lit up like a Christmas tree.

The rotor system just flew away,
I guess today just ain't my day!

1LT Brown, ORWAC Class 79-20 (Green). 64 Co., 6th Bn., 2nd Ed. Jody Calls, Fort Rucker, AL

TANK KILLER

Enemy tanks ain't so big and bad,
Rollin' out in the open, he's gonna get had.

CHORUS
*Tank killer, tank killer, lurking all the time,
Tank killer, tank killer, the enemy he will find.*

Up pops a Cobra from a coconut grove,
He's a mean machine, you could tell when he dove.
(Repeat chorus)

Lean and green and shaped like a snake,
See an enemy tank and he goes ape.
(Repeat chorus)

He's got eight TOWs strapped to his side,
If you mess with Mister snake, it's suicide.
(Repeat chorus)

CPT Jenkins, ORWAC Class 79-30 (Orange) 64th Co, 6th Bn., 2nd Ed. Jody Calls, Fort Rucker, AL

AVIATION CLASS

Aviation has a lot of class,
We don't "run" 'till we're out of gas.

Buckle up, pull th'trigger, rotors turn,
Hear that big jet turbine burn.

Pull pitch, roll in power, fly away,
Off into combat for another day.

Flyin' generals, troops, Green Berets,
Hover in, let 'em out, take it away.

Back to base, fuel it up, fly all day,
That's the way we pilots earn our pay.

Aviations has a lot of class,
We don't "run" 'till we're out of gas!

WOC Scanor, WOC Class 79-13 (Red) 61st Co., 6th Bn., 2nd Ed. Jody Calls, Fort Rucker, AL

DUSTOFF

U-H-1 on a mercy flight,
U-H-1 flying through the night.
Dustoff pilots are far and few,
Wouldn't you like to be one, too?

CHORUS
Hey, hey, all the way!
Hey, hey, every day!
Hey, hey, all the way,
Dustoff pilots every day.

Dustoff pilots live dangerously,
Flying into areas I'd rather not be.
Dustoff pilots know what to do,
I'm gonna fly dustoff, too!
(Repeat chorus)

U-H-1 hovering close to th'ground,
Hoping he doesn't take a round.
Cobras flying cover from above,
Keeping watch on the land we love.
(Repeat chorus)

Hey, hey, all the way,
Hey, hey, everyday—Aviation! Aviation!

WOC R. E. Christie, Class 78-41, 1st Ed. Jody Calls, Fort Rucker, AL

PRIMARY BLUES

I was born one morning and the sun didn't shine,
I remember my Momma was gone all the time.
The doctor looked down, and he said, "Oh, my!
Is this old boy gonna walk or fly?"

CHORUS
So you fly a fifty-five and what do you get?
Another day older and deeper in debt.
Saint Peter, don't you call me 'cause I can't go,
'Cause first I have to go and fly solo!

Now, when you see me hoverin', better step aside,
A lot of dummies didn't, and a lot of 'em died.
We got a Cyclic of iron, and a collective of steel,
If the main don't get you, then the tail blade will!
(Repeat chorus)

WOC Class 79-19 (Green), 61st Co., 6th Bn., **2nd Ed. Jody Calls**, *Fort Rucker, AL*

HUMILITY

Far above all the rest,
Aviators are the best.
True and proven by test,
Aviators above the rest.
Pilots when given any test,
Always prove to be the best.
If you doubt what I say,
Ask any pilot, any day.

64th Co. Class 78-38, **1st Ed. Jody Calls**, *Fort Rucker, AL*

AIR ASSAULT

Air Assault, Air Assault, where've you been?
Around this track and I'm goin' again.
What're you gonna do when you get back?
Jump from a plane with a hundred-pound pack.

Went to the mess hall on my knees,
Said Mess Sergeant, feed me, please.
The Mess Sergeant said with a grin,
"If ya wanna be Air Assault, ya gotta be thin!"

I got a girl who lives out west,
Thought this Air Assault life was best.
Now she's somebody else's wife,
And I'll be jumpin for the rest of my life!

I was born in a cave and raised by a bear,
I have a great big chest all covered with hair.
I have a cast iron rig cage, stainless steel balls -
I'm an Air Assault Man—and that ain't all!

CPT T. Burton, 101st Airborne Div. (Air Assault), Fort Campbell, KY

VI

Surprised to see that we march, too?
We support the rest of you!
We do P.T. to show the way,
Back to the office — every day!

Combat Support Groups

This chapter is for the Engineers, the Signalmen, Transportation, Chaplains (Yes! They do march and chant!), and all the male and female Medics with tough feet and b___s of brass.

ENGINEERS

Engineers, Engineers, can't you see?
17th Engineers is the way to be.
Paved the way through World War II,
Now we've got to follow through.

Engineers, Engineers, can't you see?
17th Engineers is the way to be.
Built a bridge across the River Rhine,
Did it all in record time.

Engineers, Engineers, can't you see?
17th Engineers is the way to be.
Fought Hitler at the Battle Bulge,
Made him wish he'd never indulged!

1LT R. Lynch, Co. E., 17th Eng. Bn., 2d Armored Div.,
1st Ed. Jody Calls, Fort Hood, TX

THE BALLAD OF EARTHMOVING PRIDE

Look out, rotor-heads, here we come,
Earthmoving Platoon's on the run.
In every project on this Post,
Earthmoving men have done the most.

CHORUS
Don't be sad and don't be blue,
Someday you might be Earthmovin', too.
We have one thing that make enemies run -
Earthmoving pride, we're Number One!

We push down trees and lay down stone,
Some nights we wonder if we'll ever get home.
When aviators fly and want to set down,
Earthmoving Platoon don't even frown.
(Repeat chorus)

We've cleared the land and all the ground,
We've built the best airstrips around.
And all you grunts, now don't forget, too,
We built that tank ditch in front of you!
(Repeat chorus)

Earthmoving gets no credit for battles won,
But we're always there before they've begin.
'dozers, scrapers, and dump trucks, too,
Scooploaders, graders, and ten-tons, new.
(Repeat chorus)

We clean up after the battle is through,
And that means Colonels and Generals, too!
(Repeat chorus)

1LT Hadley & SFC Evans, B. Co., 46th Eng. Bn., 2nd Ed. Jody Calls, Fort Rucker, AL

249 ENGINEERING

Two-Four-Ninth, Germany's pride,
Six ton truck and a bulldozer's ride.

CHORUS
Two-Four-Ninth, hey, hey, hey!
Two-Four-Ninth, every day!

We tear up roads, laugh at danger,
Weekend work's sure no stranger.
(Repeat chorus)

We build ranges, we dig ditches,
Gone so much our wives are b____es!
(Repeat chorus)

Two years at Graf, the job is done,
Back in Karlsruhe, havin' some fun.
(Repeat chorus)

249th Eng. Bn. (Combat, Heavy), Karlsruhe, Germany

C-290

C-290 roll another load,
Earthmoving man will build another road.

Shift down, throttle out, what do you hear?
C-290 is back-breaking gear.

A hard day's work and a good night's rest,
That's why earthmovers are the best!

SP5 Sullivan, C Co. 46th Eng. Bn., **2nd Ed. Jody Calls,** *Fort Rucker, AL*

DEDICATED AND DETERMINED

Way back yonder in '43,
Was born a battalion that's as good as can be.

Born on Pike's Peak in the States,
Went to Africa, but got there too late.

D-Day found them in the front,
Hittin' Normandy they bore the brunt.

Throughout Europe, they were the best,
Building bridges, longer than the rest.

Dedicated and Determined Engineers,
We sure as hell ain't got no peers!

Up in the morning at the crack of dawn,
Don't go to bed 'till the work is done.

We start each day with a few mile run,
Sure as hell is lots of fun.

Alpha Animals are mighty tough,
Be sure to holler when you've had enough.

Charlie Cobra hiss like a snake,
Damn good engineers they do make.

Delta Dawgs are known for their bite,
Sure as hell don't meet 'em in a fight.

Headquarters Heroes make the five,
Best damn engineers' companies alive!

237th Eng. Bn. (Combat), 7th Eng. Brigade, Heilbronn, Germany

MIGHTY COMBAT ENGINEERS

We build the bridges,
We lay the mine fields.

CHORUS
*We are the mighty
Combat Engineers.
We are the mighty, mighty, mighty
Combat Engineers.*

We blow the Demo,
C-4 through Bangladore.
(Repeat chorus)

We cut the L-Zs,
We can fight like Infantry!
(Repeat chorus)

Essayons! Let us try!
Essayons! Let us try!
(Repeat Chorus)

SSG M. J. Warthaw, DET A, HHC, 193rd Inf. Bde., 1st Ed. Jody Calls, Panama

ATC

Armor, Infantry, Artillery,
I gotta be in A-T-C!

Hey there, flyboy, don't you know?
A-T-C tells you where to go.

Pilots fly birds everywhere,
With A-T-C to watch and care.

Ground and tower, radar, too,
Controls your flight into the blue.

Ignore the Controller if you dare,
You will get knocked out of the air.

Take her up easy, do what you're told,
You will live longer, may even grow old.

Armor, Infantry, Artillery,
I gotta be in the A-T-C!

42nd. Co. 4th Bn., 2nd. Ed. Jody Calls, Fort Rucker, AL

WANNA BE SIGNAL

Armor, Infantry, Artillery,
One-forty-second is the place to be!

Teletypes, telephones, commos wire, too,
Don't call me, I'll call you!

SSG D. Soliz, 142nd Sig. Bn. 2nd Armored Div., 1st Ed Jody Calls, Fort Hood, TX

SIGNAL MOTTO

Ratt Rigs, Radios, Multichannels, too,
Second AD, we will support you.
Snow or rain or in the sun,
For communications, we're number one.

Grab your phone and crank the knob,
For you to talk, we do our job.
Commanders have a lot at stake,
Signal, great leaders we do make.

{Operation} was our best,
Showed the Army that we are the best!

SGT I. Roche, C Co. 142nd, Sig., 2nd Armored Div., 1st Ed. Jody Calls, Fort Hood, TX

TRANSPORTATION

We have wheels upon our collars,
We're the ones who move the Army.
Long or short, big or small,
Bring it on, we move it all.

Transportation is our name,
Moving cargo is our game.
We can go from ship to shore,
In time of peace, in time of war.

We cover every avenue,
We keep the Army on the move.
So if you ever need a ride,
Call Transportation—*we'll provide.*

SSG W.A. Honeycutt, 331st Trans. Co. AVC., Fort Story, VA

TRANSPORTATION'S CAPTAIN JACK

Hey, hey, Captain Jack!
Meet me down by the railroad track.
With my canteen in my hand,
I wanna be a Transportation man!

Hey, hey, Captain Jessel!
Meet me at dawn on my river vessel.
With my duffle bag in my hand,
I want to be a sailing man.

Hey, hey, Captain Kool!
Meet me in the motor pool.
With my girl on my arm,
I wanna be on the road by dawn.

Hey, hey, Major Hopper!
Are we gonna fly in your helicopter?
With my rifle on my knee,
I don't wanna be in the Infantry!

7th Trans. Gp. (Terminal), Fort Eustis, VA

PAC MAN

Remember my name, if you can,
I'm the one they call PAC man.
PAC man, PAC man, I'm Number One,
I fix the files but I sure can't run.

If you're in need, come to me,
I even fix coffee, but it ain't free.
PAC man, PAC man, hear my song,
I've been chair-borne far too long.

Used to train and knew how to fight,
Then "top" found I read and write.
Since that time I'm sittin' all day,
Writin' and typin' to earn my pay.
If my girlfriend asks about my death,
Tell her I was bored to my very last breath!

Anonymous, Weisbaden, Germany

FINANCE COMPANY

When the pencil is strong,
The money isn't wrong,
Chairborne, chairborne, all day.

Pay them high, pay them low,
Finance will never be slow.

We are sound, we make it right,
We are Finance, and we can fight.
Chairborne, chairborne, all day.

MSG E.D. Thornton, 502nd Finance Co. 2nd Armored Div., 1st Ed.
Jody Calls Fort Hood, TX

502nd FINANCE COMPANY

We run all the way to the cattle guards,
Everyone in Finance are P.T. stars.

Running every day makes us bold,
Keeps us away from weight control.

Running at night makes us mean,
We are the Finance Lean Machine.

Just a few words to {Enemy Leader},
From the Lean Machine of the 2nd A.D.

Hurry up, old man, make your play,
'Cause the Lean Machine is coming your way.

Move over, soldier, and let us pass,
'Cause the Finance Company is hauling cash!

SFC W.L. Teal, 502 Finance Co. Armored Div., 1st Ed. Jody Calls, Fort Hood, TX

CHAIRBORNE

Grab your typewriter and a folder,
Let's all run and get it over.

TDA is what we say,
Chairborne, chairborne, all the way.

Four battalions on this post,
But my battalion is the most!

Chapter 6: Combat Support Groups

Chairborne, chairborne, that's for me,
You can keep the Infantry.

Grab your papers, come this way,
We're the leaders—we're TDA.

Chairborne, chairborne is the best,
You dumb grunts can keep the rest.

If you read, you learn to do,
You could be chairborne, too.

TDA is what we are,
And the rest we lead by far.

Run all night, run all day,
Ain't no sweat, we're TDA.

Infantry thinks they're the best,
But 1st Battalion leads the rest.

Golden Hawk is on my shoulder,
Rosser's Raiders—none are bolder.

TDA is what we say,
Chairborne, chairborne, that's our way.

"Soldiers First," that's our motto,
And it's us—the rest must follow.

Chairborne, Chairborne, that's our desire,
Someone to run, we'd like to hire.

And in closing, let us say,
TDA—all the way!

*SFC R. Genter, 12th Co., **1st Ed. Jody Calls**, Fort Rucker, AL*

POOR ME

AG, AG, just feel free,
To send old Jody away from me.

Up in the morning, Jody's still here,
Lookin' for a handout just from me.

Run 'em, push 'em, 'till late at night,
Jody's getting stronger for the fight.

Saturday, Sunday, and holidays, too,
Jody's got your bag and has mine, too.

AG, AG, just feel free,
You gotta send Jody away from me.

SFC J. Krobath, HQ Co. USAG. XVIII Airborne Corps., Fort Bragg **Jody Calls***, Fort Bragg, NC*

AG SOLDIER

PA, PM, ASD
We are proud to be AG.
In fights so we can be free,
We support the soldier - we are the AG.

We manage the Regs,
And help with your stripes.
We keep the Army steady,
And help it win the fight.

Look upon your breast,
And what do you see?
That's the medal you earned,
Processed by the AG.

Soldier! Mark the time sheet that,
Tells us where you want to be.
Then when you get there,
It's time to thank the AG.

So AG soldier,
Wear the shield with pride.
We know that in combat,
We will all fight side-by-side.

2LT Gorriaran, AG Officer Basic Course, Fort Benjamin Harrison, IN

FIGHTIN' MAN

I joined the Army to be a fightin' man,
Now I'm in Headquarters sittin' on my can.
I shuffle papers to my left,
It's not the same as P-L-F.

I shuffle papers to my right,
It's not as exciting as a fire fight.
Air conditioning and big old fans,
I got no calloused on my hands.

My uniform's clean and my boots shine bright,
I get to sleep most every night.
Up in the morning, go to work at nine,
Get off at four 'cause I gotta' dine.

In-Box, Out-Box, what will it be?
I'm a Headquarters puke—just look at me!

PFC M. W. Lewis, 3rd Ranger Co., **Jody Calls,** *Inf. Ctr., Fort Benning, GA*

CHAPLAIN'S CHANT

C-130 goin' down the strip,
Chaplains movin' at a fast clip.

Stand up, hook up, shuffle to the door,
Waitin' to meet, meet the Lord.

First we go to Air Assault School,
Then we go to Airborne School.

Next we go to Ranger School,
Then we go to Prayer Assault School!

CPT J. P. Miller, Asst. Bde. Chaplain, 101st Abn. Div., XVIII Abn. Corps & Fort Bragg Jody Calls, Fort Campbell, KY

ARMY CLINIC

Flu shots here and flu shots there,
No more room on that derriere!
"Sick Call Blues" all day, all night,
Sing me a song while I grab a bite.

Fevers, headaches, blisters, sores,
This job's great for pacing floors.
Take a number, an aspirin, too,
Call in the morning—ain't nothin' new.

L. Patterson, Red Cross Volunteer Nurse, Patch Barracks Clinic, Stuttgart, Germany

AMBULANCE MAN

Up in the morning way too soon,
Tired as hell by the coming of noon.
427th—we're early to rise,
Performing all jobs of any size.

CHORUS
Drink a beer, damn it, drink a beer,
Well, I don't give a damn for any Old Man—
Who won't drink a beer with an Ambulance Man!

We prepared our litters, ambulances and crew,
Ready for commitments and emergencies, too.
Another mission has just begun,
The medic's job is never done.
(Repeat chorus)

Working and sweating 'til the setting sun,
While all you turkeys are havin' fun.
Hell, we're good, Hell, we're great!
If you need attention, don't hesitate.
(Repeat chorus)

We're always there to lend a hand,
If you don't take it, we'll understand.
We're always proud and certainly grand,
Never forget that we're the Ambulance Clan!
(Repeat chorus)

427th Medic. Co., 2nd Ed. Jody Calls, Fort Rucker, AL

601st MEDICAL COMPANY

I went to me genie for some advice,
And what she said didn't sound so nice.
She got a letter from Uncle Sam,
That said I'd go to Vietnam!

So I grabbed my boots and I hit the trail,
It was either that or go to jail.
Oh, Uncle Sam, he's pretty mean,
He sent me off to do my thing.

He put an 'aid bag in my hand,
And sent me to places I'd never been.
The fightin' is over and the battle is won,
But the medic's job is never done.

Oh, General, General, come see us,
We know you're busy—but it's a must!
We're the best darned Company you ever did see -
We're the 601st Medical Company!

SPC Cheers, 601st Med. Co., 1st Ed. Jody Calls, Fort Clayton, Panama

MESS HALL

I joined the Army to keep the peace,
What am I doing frying chow in grease?
I'm the victim of jokes, people think it's fun -
But they eat that food when the laughin's done.

Mess Hall, 66th MP Co., Karlsruhe, Germany

94B

Oh, Recruiter, can't you see?
I wanna be in the Army!
Yes, be all I can be!

Put me where I'm needed most,
I don't care which Army post.

So you want to be needed, the Recruiter said.
"Well, there's thousands of troops who need to be fed.

If you wanna be all that you can be,
I'm gonna make you a 94B!"

Here at Fort Benning, the moon shines bright,
But I gotta take leave to see sunlight.

I wake up in the morning, most times at three,
To a thousand troops saying, "Feed me, PFC!"

There's not enough grits on the old chow line,
The troops are here, head counts not on time.

So if you wanna be all you can be—
Try the life of a 94B!

PFC P. A. Wilson, Co. C, 3rd Bn., 1st ITB Jody Calls, Fort Benning, GA.

MPs

We're MPs, see how we roar,
Clean up the street, then beg for more.
Throw out your pot, dump your beer,
Better look sharp, MPs are near!

Use your signal, stop at signs,
Don't drive too fast, toe that line.
Fight with your spouse and we'll be there,
When you do something stupid, better beware.

MPs, Fort McClellan, AL

VII

Look what's happened the past twenty years,
These had to be censored for delicate ears.
'Nam is a war some would rather forget,
But it's part of the past for every Vet.

Desert Storm, Vietnam War, and Current Trends

This chapter taps into lots of conflicts from Vietnam to Iran, and from the Middle East to Grenada. Many chants, of course, still hold the old Soviet-style Russia as a deadly enemy; the hatred for Red Communism runs deep. Even though these chants may appear a bit out-dated, they are a serious and proud part of our Cold War history when American Armed Forces held the line until Soviet Communism shriveled up and died.

It's important that the American military keeps alive the fighting spirit for which it is famous. You can do your part to make it live on by singing these chants out either "as is," or by punching into the right slots the names of the more current enemies to freedom and justice. Let these grand old chants ring out around posts and shipyards and beaches and airfields; let's keep these traditional American battle cries living on and on and on!

MAKES ME FEEL SO GOOD!

(Leader)
    Tell me what I say!
(Troops)
    Makes me feel so good!
(Leader)
    I can run all day,
(Troops)
    Makes me feel so good!
(Leader)
    I can fight all night,
(Troops)
    Makes me feel so good!
(Leader)
    I can run all day, I can fight all night,
(Troops)
    Makes me feel so good!
(Leader)
    Well, I'm Infantry,
(Troops)
    Makes me feel so good!
(Leader)
    Come on and follow me,
(Troops)
    Makes me feel so good!
(Leader)
    I'm Infantry, come on and follow me,
(Troops)
    Makes me feel so good!
(Leader)
    Crossed rifles on my chest,
(Troops)
    Makes me feel so good!

(Leader)
> Makes me America's best,

(Troops)
> Makes me feel so good!

(Leader)
> Crossed rifles on my chest, makes me America's best,

(Troops)
> Makes me feel so good!

(Leader)
> Let the Commies come,

(Troops)
> Makes me feel so good!

(Leader)
> And I'll make 'em run,

(Troops)
> Makes me feel so good!

(Leader)
> Let the Commies come, and I'll make 'em run,

(Troops)
> Makes me feel so good!

(Leader)
> I'll fight on Cuba's shore,

(Troops)
> Makes me feel so good!

(Leader)
> Or El Salvador,

(Troops)
> Makes me feel so good!

(Leader)
> Tell me what I say!

(Troops)
> Makes me feel so good, so good, so good!

1LT B. C. Reynolds, 3/5 Inf., 193rd Inf. Bde., Panama

TOUGH AND MEAN!

This U.S. soldier is mean and tough,
The Ayatollah's (Saddam Hussein's) gonna have it rough.

Big man, small man, shoot that gun,
Keep Khomeini (Hussein) on the run.

Anonymous

COLORING LESSON

Color our troops red, white and blue,
Ever loyal, ever true.

Color Saddam a bright, bright yellow,
He talks big for a chicken-fellow.

Color all Commies a deep, deep red,
We'll chase 'em down 'til they're all dead.

Color our Army a dark, dark green,
A fighting machine, quick and lean.

Anonymous, Fort Richardson, AK

GOT TO MAKE OUR PRESENCE FELT

Shoot my fifty into space,
Hope it hits a Russian (an Iraqi?) face.
Lock and load my forty-five,
Try to kill every Commie (Iraqi) alive.

Aim my M-16 down range,
If Ivan (Saddam) survives, it'll be real strange.
Empty my sixty of every belt,
Got to make our presence felt!

SPC P. Kelley, D. Btry, 1/41 FA 56th FA Bde., (Pershing), Germany

AIRBORNE RANGER

I wanna be an Airborne Ranger,
I wanna live a life of danger.
I wanna go to Vietnam,
I wanna kill some Charlie Cong.

I wanna be an Airborne Ranger,
I wanna live a life of danger.
I wanna go to Israel,
I wanna raise all kinds of hell!

Anonymous

Note: This particular cadence was one of the most popular to emerge from the Vietnam War. The unrest in the Middle East added the second stanza. What will be added next?

SLIPPERY SAM

Up in the jungle of Vietnam,
Came a Recon Marine they called Slippery Sam.
He wore a string of ears right across his chest,
Just to show Charlie he was always the best.

Ten, twenty, thirty, forty, fifty or more,
Sam kept shootin' them and addin' up the score.
Many V.C. died trying' to kill this Marine,
But Slippery Sam was too darned mean.

One day while crawling through the jungle trees,
Sam shot a gook right in the knees,
Sam pulled out his K-Bar before he died,
And stuck it right between his eyes.

One day on a hill they called Khe Sanh,
Sam decided to have some fun.
He put fifty claymores in a line
 And then watched Charlie blow his mind.

MAJ Wm. Windsor, U.S. Marine Corps, (Ret.)

CARTER RUNS

Raise your eyes and what do you see?
Mister President is running with me.
2nd I.D. is above the rest,
Mister President runs only with the best.
Casey, Casey, second to none,
Look at us and Jimmy Carter run.

Anonymous

IN VIETNAM

Saigon Sally's cryin'
'Cause all her boys are dyin'
You know the Infantry is here,
In Vietnam ... In Vietnam.

Late at night while you're sleepin'
Charlie Cong comes a-creepin'
You're runnin' out of ammo;
you've fired your forty-five,
You pray your hand-to-hand, will keep you alive.
In Vietnam ... In Vietnam

J. H. Lutz, Toronto, Canada

CHARLIE

(Sung to the tune of "C-130")
Crawling low and feeling mean,
Spot Charlie troops sittin' by a stream.
Attack at once, listen to 'em scream,
Shoot them up with my M-16.

There is Charlie in his boat by the stern,
He don't think his boat will burn.
Old fool Charlie has a lot to learn,
Charlie meets the napalm burn!

PFC C. D. Nelson, B. Co., 2nd Bn., 30th Inf., Ledward Barracks, Schweinfurt, Germany

WHAT THE RANGERS DID TO ME

See my buddies on the ground,
Hear those bullets whizzin' round.
They gave me a knife, taught me to fight,
Sent me out in the middle of the night.

CHORUS
Mama, Mama, can you see?
What the Rangers did to me?
Mama, Mama, can you see?
What the Rangers did to me?

Saw the sentry, brought him down,
Felt the blood come oozing 'round.
Taught me how to kill a man,
Now I've had to—I know I can.
(Repeat chorus)

Hear me wake in the middle of the night,
Sweatin' and shakin', remembering the fight.
The day may come, I hope it can,
When I won't need my knife again.
(Repeat chorus)

But Mama, 'til that day arrives,
I'll practice taking other's lives.
When you ask why that must be,
One big word will describe me -

Ranger! Mama, that's what I am,
One of a breed of fightin' men!

A. L. Brown, Anchorage, AK

WE WON'T FORGET

Russia, Russia, feeling blue,
'Cause we got the goods on you.
Shooting down planes from outta the sky,
Sending Commies to kill and spy.
We may not be at war yet,
But you done wrong; we won't forget!

Anonymous, Fort Bliss, TX

COMMIE

Look on the hill, what do you see?
Dirty Russian Commie, lookin' at me.
Lock and load one round inside,
Shoot that Russian between the eyes.

Beat and kick and stomp his face,
Then strangle him with your boot lace.
And if he's still a fightin' fool,
Do him in with your entrenchin' tool!

2LT Jesus Huerta, IOBC, Fort Benning, GA

VIETNAM

I hear the choppers hovering,
They're hovering overhead.
They've come to get our wounded,
They've come to get our dead.

CHORUS
Alpha kill, Alpha kill,
Late at night when you're sleepin,
Alpha Company comes creepin' all around.

I hear the mortar's volley,
The shell are dropping all around.
I see my buddies falling,
They're falling all around.
(Repeat chorus)

I see my buddy's wounded,
He's shot between the eyes.
I say he's gonna make it,
He turns around and dies.
(Repeat chorus)

I see our flay a-wavin,
It's wavin' overhead.
It's wavin' for our wounded,
It's wavin' for our dead.
(Repeat chorus)

SGT R. A. Draper, MP School, Co 11. M.P.T. Bn., Fort McClellen, AL

AIRBORNE FLIGHT

Grenada was a proper fight,
For those of us in Airborne flight.
We left there with our heads held high,
Airborne soldiers never die.

Freedom's just another word,
If you're just an average nerd.
But when you are Airborne tough,
Foes will know you never slough.

Enemies who hear this sound,
Have always to their regret found.
We will always ruin their day,
We are Airborne all the way.

When we face the Judgment Day,
We will never shy away.
We always hold God's foes at bay,
We are Airborne all the way!

MAJ A. Gonzales, Human Resources Mgt., **Jody Calls, XVIII Airborne Corps and Fort Bragg, NC, Fort Campbell, KY**

RANGER, RANGER

Ranger, Ranger, on the line,
Gotta make a jump just-a one more time.
Down below the jungle lies,
Ranger's got a tear in his eyes.

CHORUS
Sayin' every-day, All-the-way!
Super-duper, Paratrooper,
Rock-steady!

Hits the ground on the go,
Will he live? He don't know.
Gotta hit that Russian (Iraqi) base,
Put in that Commie's (Saddam's) face.
(Repeat chorus)

Parachute flare pops over the site,
Ranger, Ranger starts the fight.
One by one the Commies (Iraqis) fall,
One more minute and he's got 'em all.
(Repeat chorus)

Star Cluster signals the end of the fight,
Ranger Patrol slips into the night.
Back at the site the Ranger lies,
Clutching for his 16, he slowly dies.
(Repeat chorus)

Back at home, his widow cries,
DSC just ain't no prize.
Her Ranger's dead and that's a fact,
He's just another Ranger that ain't comin' back.
(Repeat chorus)

1LT B. J. Orloff, A. Co. 1st Bn. 4th Inf., Aschaffenburg, Germany

CHARLIE

(Sung to the tune of *C-130*)
Crawling low and feeling mean,
Spot Charlie troops sittin' by a stream.
Attack at once, listen to 'em scream,
Shoot them up with my M-16.

There is Charlie in his boat by the stern,
He don't think his boat will burn.
Old fool Charlie has a lot to learn,
Charlie meets the napalm burn!

PFC C. D. Nelson, B. Co., 2nd Bn., 30th Inf., Ledward Barracks, Schweinfurt, Germany

NURSE STORY

Our Chief Nurse is a wide old man,
Told us Saudi sun makes a perfect tan!
We all paid our dues.

Complaining 'bout the sand, no beer!
Those well-hidden fears.
We all paid our dues.

Baghdad Betty's threats at night,
Only stopping in the light.
We all paid our dues.

Scorpions hide in desert boots,
Hair straight with darkened roots.
We all paid our dues.

150th Field Hospital, Army Reserves (Desert Shield/Storm)

GOING TO IRAN

I am going over to Iran,
With a rifle in my hand.
And when I'm finished over in Iran,
I am going to Afghanistan.

We'll move the nasty Russians out,
We're gonna move them without a doubt.
Kicking and fighting and cutting all day,
We don't know no other way.

Drill Instructor Chant Book, 1/85 U. S. Marine Corps Recruit Depot, San Diego, CA

KOREAN PT RUNNING CHANT

One mile, no sweat.
Two miles, can do.
Three miles, will do.
Four miles, so good.
Five miles, every day.
Kimchi - (hand clap!)
Mockli - (hand clap!)
Bulgogi - (hand clap!)
So good, so good! - (hand clap!)
Every day, all day - (hand clap!)
Every night, want to fight - (hand clap!)
All the way, every day!

SSG J. W. Herber, 225th Inf., Michigan Nat'l Guard

STREET SMART

The Crips and Bloods could learn from us,
Stabbing and shooting, ready to cuss.

We do what we're trained to do,
Fighting for freedom for me and you.

Street gangs stealing and shooting up drugs,
We know how to teach those thugs.

Use those muscles on Mister Hussein,
Show him street-smart is the name of the game.

Anonymous, Fort Lewis, WA

IN THE MOVIES

Movie stars have nothing on us,
Our moves are made without that fuss.
Panama was a practice run,
Jumpin' in Grenada was just for fun.

Rambo could learn a thing or two,
We left him behind, nursin' his flu.
Cameras ready, crew on call,
We're marchin by, standing tall!

Anonymous, Air Assault School, Fort Campbell, KY

CENTRAL AMERICA

Rumors about earning hazardous duty pay,
Way down Central America way.
Well, I don't like rice but I like beans,
And I don't like that jungle green.

You can't tell who's friend or foe,
Seems everyone wants to kill G.I. Joe.
My Spanish skills are mighty rusty,
Already my BDUs are all crusty.

Fightin' men go where they're sent,
Blood and guts help pay the rent.
I hear it's like another 'Nam,
Just as hot but not so crammed.

If Reagan decides to pull the cord,
We'll all pray harder to the Lord.
We'll need faith more than ever before,
If Central America gets into war.

Anonymous, Fort Bliss, TX

AYATOLLAH KHOMENI

(Sung to the tune of *C-130*)

Ayatollah Khomeni,
What'cha gonna do with our Embassy?
America is a peaceful land,
But when we're pushed we'll make a stand.

Ayatollah, you old fool,
You have gone and lost your cool.
Land of the free and home of the brave,
We're gonna put you in your grave.

Bomb Iran, bomb Iran,
If they're not careful, we're taking that land.
Nuke 'em, nuke 'em 'til they glow,
If they don't let our people go!

Iran, Iran Parliment,
You haven't even made a dent.
Free your hostages today,
Or we will have to earn our pay.

*2LT R. Brown III., A Co. 3/66 Armor 2nd Armored Div.,
1st Ed. Jody Calls, Fort Hood, TX*

GRENADA ISLAND

Grenada Island, here I come,
To save you from the Commie scum.
Pack on my back and 'chute on my chest,
I'm plenty scared. but I'll do my best.

C-130 at 500 feet,
'Chute opened up and I hit the street.
Saved some students, set 'em free,
Tied those Cubans to a tree.

VICTORY soldiers, we're the best!
What little island will be the next?

1LT T. Myers, HHC, 2/70 AR, 24th Inf., Fort Stewart, GA

IN THE EARLY MORNING RAIN

It's our first week of training,
All around it's raining.
But they won't stop our training,
In the early morning rain.

Sergeants make us run and fight,
All the day and through the night.
As they smile with delight,
In the early morning rain.

CHORUS
In the early morning rain,
With my weapon in my hand.
And a pocket full of sand,
In the early morning rain.

All around the war is near,
And it's war we all fear.
As we drink our last can of beer,
In the early morning rain.

Got the enemy to my front,
And the ocean to my rear.
Wounded and dying's all I hear,
In the early morning rain.
(Repeat chorus)

As I'm lying here to rest,
Catch a bullet in my chest.
Even though I've done my best.
In the early morning rain.

Even though I'm gonna die,
Tell my darling not to cry.
'Cause I'll never say goodbye,
In the early morning rain.
(Repeat chorus)

Many a solider will die today,
Guess there's nothin left to say.
So our children they can play,
In the early morning rain.

Yes, my Sergeant I can NOW see,
Why all this training's good for me.
Forever more we'll be free,
In the early morning rain.
(Repeat chorus)

*SFC R. Dunlap, Drill Sergeant, Co. E. 11th MP Bn.,
Fort McClellan, AL*

NEW ENEMY

There's something out there give me the scare,
Worse than Charlie or a Russian bear.
You can't see it until it's too late,
I can't think of a more terrible fate.

I'm not on drugs and I'm not gay!
But they tell me I could die today.
The short name of this mean old germ,
Is one that makes us soldiers squirm.

Stick to your wife and don't play around,
Or AIDS will put you in the ground!

Anonymous, Fort Leonard Wood, MO

2nd ID

It's hard to be so far from home,
Carrying this gun in the DMZ zone.
We're soldiers proud, soldiers brave,
Our country and loved ones we're here to save.

Watching for tunnels and men in the sky,
Ready for action, willing to die.
At night voices coming out of the dark,
Those North Korean goons think it's a lark.

Anonymous, 2nd Inf. Div., Camp Casey, Korea

HEALTH & WELFARE

Health and Welfare in the night,
Sit on your bed and turn on the light.
Look for women and look for drugs,
Look in the ceiling and under the rugs.

We find women, whiskey, too,
Hope we don't find any with you.
Always remember as you play,
The Company Commander can cut your pay.

Bring the dogs!
Bring the dogs!
Whoof!
Whoof! Whoof!
Whoof!
Whoof! Whoof!

CPT T. Fultz, HHC, 237th Eng. Bn., Germany

I AM AIR DEFENSE ARTILLERY

Air Defense Artillery,
Patch on my shoulder.
Jump on your Vulcan and follow me,
I am Air Defense Artillery!

If I go to Singapore,
Tell 'em what I went there for.
And if I go to Chinatown,
Tell 'em I don't mess around.

Air Defense Artillery,
Patch on my shoulder.
Jump on your Chap and follow me,
I am Air Defense Artillery!

If I go to Iran,
Tell 'em that I got my man.
And if I go to Vietnam,
Tell 'em that I love my Mom!

Air Defense Artillery,
Patch on my shoulder.
Jump on your FAAR and follow me,
I am Air Defense Artillery!

1LT S. W. Poorman, D. Battery, 2/5 ADA 2nd Armored Div., Fort Hood, TX

HURRY UP! DON'T DELAY; SEND YOUR BEST JODY CALLS IN TODAY!

Mail in your Jody Calls to:

**JODY CALL EDITORS
THE TALISMAN MEDIA GROUP, INC.
336 BON AIR CENTER, SUITE 341
GREENBRAE, CALIFORNIA 94904
(415) 461-6409**

JODY CALL NOTES

*Submission shall not constitute a contract to publish. The Editors reserve the right to reject any submission without grounds. No cadence will be published that uses offensive terms or is deemed by the Editors to be not in the best interests of the United States government or its Armed Forces.

IT'S *YOUR* TURN TO SOUND OFF!
Be a Part of Military History!

FACTS:

☞ *ONLY YOU* can help us keep the tradition of Jody cadences alive!

☞ *ONLY YOU* can help make your own unit or branch immortal!

☞ *ONLY YOU* can help the Talisman Books compile even more volumes of the new Jodies your chants while marching and running.

☙ Sooo, write 'em down, send 'em in, and we'll probably print 'em.*

Here's how YOU can make your contribution to keep the best U.S. military Jody Chants alive:

- Write your Jodies down <u>exactly</u> as they are chanted.
- Make sure they are complete in every verse and chorus.
- Send them in along with your name, rank or rate, unit and service so we can give you the credit you deserve.

YOUR REWARDS:

If the Editors select your Jody Chant to be included in a future publication, we will send you your own *free* copy hot off the presses. Use it, show it to your buddies, send it home to your special friend or your Mom, whatever. It's yours to keep—and it will be autographed to you personally by the Editors!

AND MORE!

At their sole descretion, the Editors will select one outstanding Jody Chant from each branch of the U.S. Armed Forces as the *"most likely to be remembered through the years."* If your Jody is selected, you will receive <u>five</u> more autographed copies of our very next *Jody Handbook* in which your chant is included; PLUS a Letter of Appreciation from the Editors suitable for framing; PLUS a certified check for Fifty Dollars ($50.00)! ☞